The Baseline Concept in Biodiversity Conservation

Series Editor
Françoise Gaill

The Baseline Concept in Biodiversity Conservation

Being Nostalgic or Not in the Anthropocene Era

Edited by

Laurent Godet
Simon Dufour
Anne-Julia Rollet

iSTE

WILEY

First published 2022 in Great Britain and the United States by ISTE Ltd and John Wiley & Sons, Inc.

ISTE Ltd
27-37 St George's Road
London SW19 4EU
UK

www.iste.co.uk

John Wiley & Sons, Inc.
111 River Street
Hoboken, NJ 07030
USA

www.wiley.com

Library of Congress Control Number: 2022941253

British Library Cataloguing-in-Publication Data
A CIP record for this book is available from the British Library
ISBN 978-1-78630-888-7

Contents

Introduction

I light a fire in the Pyrenean hut after having searched all day, in vain, for the brown bear in the beech–fir forests with their mossy soils. The beast was, however, close to me, the photographic traps placed by the professionals and amateurs of the "*réseau ours*" would reveal it to me a few days later. It does not matter. What I came for was instead a break from the rest of the world. A "different place", a little escape from my fellow kind. So, when I see the tracks of an Pyrenean chamois hunter and I glimpse the silhouette of a hiker between two silver fir trees, when I hear the tinkling of the last cowbells of a herd returning or, worse, the horn of a car down in the valley, I pretend not to have perceived anything as it disturbs an almost perfect atmosphere. Because when I arrived at the hut at nightfall, it is fair to say that I was hoping that no candle or headlight was even there.

It is almost shameful to write these lines in a book or say them in a lecture hall as a scientist working in the world of so-called "conservation science", for several reasons.

The first is that to go looking for bears seems implausible to many in 2021. For most people, this species no longer belongs to French fauna. It is a species that populated French forests thousands of years ago, which still exists in the forests of Siberia or in the great parks of the American West, or, even worse, as an exotic species that was introduced ex nihilo into our mountain forests from Slovenian populations. In any case, there are none or none left. And even if your interlocutor were to believe you, they would ask you in the following second if it is not dangerous to be face to face with the plantigrade. Let us recall the facts: there were at least 52 bears in the Pyrenees in 2019 (Sentilles *et al.* 2020), it is a Holarctic species, therefore also European (Wilson and Mittermeier 2009), which has never disappeared from French soil (Etienne and Lauzet 2009). Even better, it is a totemic

---

Chapter written by Laurent GODET.

species, a king of animals in Europe for a long time. Pastoureau (2015) reminds us of the existence of mixed Neolithic burials between bears and humans, the presence of bears in the genealogies of Danish kings or the numerous pagan rituals dedicated to the beast (which were also the subject of violent attacks by the Christian Church). Concerning the dangerousness of the species, even if encounters with the animal make the headlines in the regional newspapers, there have been no fatal attacks on humans since the reinforcement of populations in the Pyrenees in the 1990s. There are only reports of intimidation charges from females with cubs. There is therefore a lack of knowledge of the history of the living world. Many know how to date the great fire of London, the tsunami of Lisbon, the fall of the Western Roman Empire, because they have learned, revised and crammed in their books. But how many people know that the bison, the elk or the aurochs were still present on French territory in the 5th–7th, 10th and 12th centuries, respectively (Pascal *et al.* 2006)? It is even a question of what can be called collective amnesia, documented and theorized as a "shifting baseline syndrome" (Pauly 1995). In just a few generations, we forget what the history of our environment was, taking as a baseline our childhood or, at best, that of our parents or the story of our grandparents. We forget and we get used to it. At most, we are nostalgic for a very recent past. How then can we imagine a return of the living in our societies if they forget that certain species existed alongside them so little time ago? The gray wolf, for example, populated France throughout the Holocene, but its disappearance during a very short time window of 70 years, from the 1920s to the early 1990s (Moriceau 2011), made coexistence with the animal even more unbearable for many, to the point that falsifications of the history of the wolf's return were invented from scratch to justify the policy of eradicating the species in France[1].

The second reason is the sensitivity of the words of this introduction. For many people, science and sensitivity do not always seem to go together. It is difficult to understand that a scientist studies an animal or works for its conservation because they have a sensitive relationship with it. To put it simply: they love the species they are working on. Imagining that in a given place or at a given time, this species is or was present constitutes a comforting spatial or temporal baseline. But one expects rational arguments to legitimize the protection of living things or, quite simply, to explain the choice of one's study models. A student who wishes to work on whales or great apes will very quickly be qualified by those around them as a "dreamer". Being transported, animated and passionate is not acceptable. The progressive shift from "protection" to "conservation" (like the passage from the International Union for the Protection of Nature (IUPN) to the International Union for Conservation of Nature (IUCN)), or from "nature" to "biodiversity", corresponds more or less to the

---

1 Including by politicians like Christian Estrosi, claiming that the species had been "artificially reintroduced" (sic) by state officials and park rangers in Mercantour Park. His conviction for defamation in 2015 was then overturned on appeal two years later.

transmission of a great whole to the quantification of elements distinct from one another. Protecting nature for its intrinsic value has become a difficult position to hold. It is more easily protected using functional arguments (because nature provides functions) and instrumental arguments (because nature provides services to human societies[2]). The few great founding narratives about nature and the wonder of being in nature (*e.g.*, Thoreau 1854) have thus been undermined by several authors[3]. It would therefore no longer be acceptable to want to protect nature on the grounds that we love it and are amazed by it. It would be even less acceptable to forge a sensitive image of it because, for many, it is necessary to face the evidence: nature would not exist any longer.

This is the third reason that discredits the proposition of a "quest for the bear" (to use Salingue's (2015) formula) in France in 2021. Nature would no longer exist. The bears I could have seen are reintroduced individuals (originating from Slovenia) or individuals from reintroduced parents. The one that was photographed near me a few days before my visit even has a name. We know the female with which he mated and we even know his offspring whose names were given by the children of the villages of the valley. The forests that I walked through are partly planted by humans and have been cut down. The hut in which I slept is an artifact and is located in the middle of a mountain pasture that exists only because of pastoralism that goes back several centuries. My ascent was along a long-distance hiking trail that is used heavily in the summer and used as a cross-country ski trail in the winter, to the point that the plant communities have been modified. I hid in a clearing that is nothing more than a firebreak maintained for silvicultural purposes. Even the marmot that screamed when a golden eagle circled above me is a species introduced to the Pyrenees in the 20th century (Pascal *et al.* 2006) to give these mountains the image of the chic resorts of the Alps.

As a result, entire disciplines now balk at talking about nature. As Maris (2018) points out, some may make a "fatal prognosis about the end of nature". There would not even be nature or wilderness anymore. Works with titles evoking a post-nature or post-wilderness world have followed one another in recent years. Meyer (2006) speaks of the "end of the wild", Marris (2011) of "post-wild" and Lorimer (2015) of "conservation after nature". French geographers, on the contrary, have only used the term "nature" or "natural" with quotation marks for a long time (see, for example, Demangeot (1984)). The dissociation between nature and artifice thus seems to be a

2 For a critical approach to ecosystem services, see Maris (2014).

3 Peter Kareiva, one of the most fervent representatives of the New Conservationists, strives, for example, to point out inconsistencies in Henry David Thoreau's story. In a style mixing arrogance and contempt, he tries to underline all the mawkishness that there would be in the account of the person who cut themselves off from the city by staying in a cabin on the edge of Walden Pond in the middle of the 19th century.

thing of the past. This form of relativism puts everything on the same level, on the grounds that everything bears more or less the mark of humans.

This last point makes me dizzy. I wish to extract myself from the civilized world and, in doing so, I pass for a non-conformist one with an idealized vision of nature in a society that has obviously cut its ties with it for generations. Indeed, nothing is really natural anymore, except the idea I have of it. But should we stop thinking about this? Is it desirable, joyful or, on the contrary, frightening to find ourselves in this Anthropocene? Is the entry into the human era a journey on a train with no return, launched at full speed, where the only baseline would be the present cut off from the past and with a future as uncertain and unpleasant as the continuation of the sixth extinction of life? Could the Anthropocene be this "terrible failure of humanity" (Descola 2015) or this "one-way trip for humanity to an uncertain future" (Steffen *et al.* 2011)? Faced with such an observation, the individual may feel helpless (see Cazalis and Granon (2017)), or even plunged into a cynical fatalism by accepting to live in a situation of danger (Beck (1992, 1994) cited in Eckersley (2017)). In a nutshell: What can we do with past baselines in a context of the crisis of the living and increasing domination of humans over the planet?

The first step proposed in this book is to consider the Anthropocene not as a brutal, definitive and indisputable event, but rather as a heuristic metaphor, like Lovelock's Gaia hypothesis (1979). Above all, it is a concept that helps to illustrate the unprecedented hold of humans on the Earth system. Then, it is a question of discussing the diversity of responses that are being implemented in order to loosen this grip (*e.g.*, conservation, restoration, rehabilitation or compensation) and more precisely a key concept that, explicitly or implicitly, guides these responses: baseline. Thus, rather than remain appalled by the "Anthropocene statement", this book proposes, on the contrary, to analyze this concept in order to have a critical approach through three parts.

Is the hold of humans monolithic? Monolithic in space first. Social scientists thus remind us that many societies have had little or no involvement in the changes of the Anthropocene (see Lowenhaupt Tsing 2017; Malme 2017; Glowczewski and Laurens 2018) so much so that some authors have preferred to speak of the "Capitalocene" (Haraway 2015; Bonneuil 2017), the "Technocene" (Hornborg 2015), or the "Anglocene"[4] (Fressoz 2017). Monolithic in time then. Did the Anthropocene arrive very suddenly? Bonneuil and Fressoz (2013) thus remind us that the "Anthropocene event" itself and its conceptualization are not recent. Therefore, can we identify baselines as tipping points in time during which changes in the hold of humans on the Earth took place?

4 In the 20th century, the United Kingdom and the United States alone accounted for more than half of greenhouse gas emissions until very recently (Fressoz 2017).

What do we want to do once we have identified possible baselines? Are these changes necessarily irreversible? Can they be stopped? Are there no possible returns to past baselines or should we, on the contrary, let nature take its course, or even welcome the arrival of hybrid "new ecosystems"?

Finally, how can these reflections be translated into practice regarding concrete cases? Using the example of the coastline, forests and the countryside, we try to understand how this notion of baseline can be mobilized.

The aim of this book is to offer geographers' multidisciplinary views on the concept of baseline. Indeed, the latter questions the relationship that societies have simultaneously with temporal dynamics, spatial diversity and the way of living in nature, all themes of reflection embraced by the geography of nature.

PART 1

Defining Baselines

1

Temporal Baselines: Finding a Tipping Point in the Past

1.1. Preamble

Leaving Paris for Mansle, a small village in the Charente region of France, Rachel drove across France on a hot July day. Her two children were sleeping in the back seat. They fell asleep after a few minutes, probably tired from the monotony of the Beauce countryside. They were not thrilled by the sight of this French granary and biological desert, where only a few intrepid Kestrels brightened up the landscape every 5 or 10 km with their Flight of the Holy Spirit over invisible prey. Four hours later, arriving in the courtyard of her elderly retired parents' Charentais farm, Rachel saw her mother who exclaimed:

Did you wash the car in Ruffec? The windshield is all clean!

The children did not understand their grandmother's remark. She quickly explained to them that in the past, when we drove in the summer, we had to clean our windshields, because there were swarms of insects that crashed into them. All of this vaguely reminded Rachel of something and she questioned her mother:

Do you know when exactly? I mean... do you know when we stopped having these bugs coming crashing into cars?

I don't remember, but it must have been before we sprayed everything, replied the grandmother.

Chapter written by Laurent GODET, Simon DUFOUR, Anne-Julia ROLLET and Armelle DECAULNE.

Sprayed everything? asked Rachel

Yes, before we put all the chemicals in the fields. Now, even when Thierry harvests, it's only dust, no bugs. There are not even grasshoppers anymore!

The swarms of insects that crash into cars and spring up from the wheat at harvest time seem to belong to another time, the grandmother's apparently, not really that of the mother. As for the children, it all belongs to science fiction. Clément, five years old, imagined that his grandmother's childhood probably coincided with the beginning of the extinction of the dinosaurs. But all this raised questions for Rachel who mumbled softly:

When exactly did we stop having insects and therefore birds and other animals? Was it the same as global warming? Was it at the same time?

Rachel did not get an answer from her mother, who was obviously busier swatting a fly that had landed on the oilcloth of the kitchen table. Looking away, she wondered when everything changed. What if "it was better before" was not just a formula passed down from generation to generation? What if there really was a moment, a clearly identifiable period, when humans began to destroy everything?

1.2. Introduction

Since the first classifications of living beings by Carl von Linnaeus (18th century), humans have placed themselves above other species. Since Darwin, we have known that this place has been usurped, *Homo sapiens* being only a member of the great community of living beings among the two million or so species that he strove to describe and classify. In this context, is it just as pretentious to attribute to *Homo sapiens*, the name of a new period of the age of the Earth: the Anthropocene?

At first sight, everything alines to say that it is not. What makes our species original is not the fact that it is more intelligent, "more evolved", that it speaks, has a sense of humor or even plays. What makes our species original is its unique ability to modify the Earth system as a whole, to the point that our species is a major force that is profoundly modifying the atmosphere, the lithosphere, the hydrosphere, the cryosphere or the biosphere (see, for example, Harari (2015) for a general overview). Regarding the atmosphere, humans have emitted 555 petagrams of carbon into the atmosphere since 1750, generating a concentration of this gas not seen in at least 800,000 years (Lewis and Maslin 2015). This concentration of CO_2 has led to ocean acidification, also not seen in at least 300 million years (Lewis and Maslin 2018). Successive IPCC reports confirm that the emission of all greenhouse gases released by humans is changing the Earth's climate, leading to a rise in temperature and sea levels. The consequences on the cryosphere are major: the melting of the Greenland ice cap (with a loss of more than 200 gigatons of ice per year (GIEC 2013)) and that of Antarctica (50 gigatons per year (GIEC 2013)) are striking evidence of this warming. From a biosphere perspective, species extinction rates are 100–1,000 times higher than historical extinction rates (Barnosky *et al.* 2011) and the globalization of trade has led to a great mixing of species across the globe and biotic homogenization (Baiser *et al.* 2012).

As Magny (2021) reminds us in his excellent synthesis on the Anthropocene, this capacity of humans to modify the entire planet they inhabit was identified very early on. At the end of the 18th century, Buffon, in *The Epochs of Nature*, identified seven epochs, the seventh being the one in which the power of humans assisted that of nature. Jenkyn, an English clergyman, spoke of *anthropozoic* as early as 1854, a term taken up by the Italian geologist and priest Stoppani in 1873, while Vernardsky, Theilhard de Chardin and Le Roy spoke of the *noosphere*, as, according to Vernardsky, the new phase of the Earth where human cognition transforms the entire biosphere (Magny 2021). Several authors questioned, however, whether humans have a sufficiently marked and identifiable impact to speak of an obvious geological time unit. According to Lewis and Maslin (2015), however, the Anthropocene must be defined on the same criteria as all the previous eras, because otherwise the Anthropocene would be merely an ideology.

In this work, which deals with the notion of baseline and conservation of nature, the question is therefore raised not only about the relevance and possibility of identifying the beginning of a human hold on the Earth, but also the various dates that can be considered as the "tipping point(s)" from which this hold clearly began. In parallel to the anthropogenic dimension as a factor of change, the temporal baseline also questions the existence of a partial reversibility of the functioning of ecosystems or their functionality.

1.3. Recognition problem: how do we define a new unit of time?

The geological time scale is composed of different chronostratigraphic units (corresponding to sedimentary layers of the Earth's crust) combined with geochronological units (corresponding to time intervals). The geological time scale was standardized by the International Commission on Stratigraphy in 1974. Geochronological units are eons, eras, periods, epochs (see Figure 1.1) and stages, each with opposing chronostratigraphic intervals (eonothems, erathems, systems, series, stages and substages) (see Ogg (2004); Gradstein *et al.* (2012)). All units of the Phanerozoic Eon (542 million years ago to the present day) and the last Proterozoic period (Ediacaran, 632–542 million years ago) are defined by a Global Boundary Stratotype Section and Point (GSSP); commonly referred to as the golden spike on the charts; the Archean and Proterozoic eons (*i.e.*, the entire Precambrian interval excluding the Ediacaran), prior to 632 million years ago, are subdivided into absolute ages (Global Standard Stratigraphic Age or GSSA) due to the absence of fossils. The golden spikes, therefore, define a specific point at the lower limit of a geological stratum, which must be caught between two golden spikes.

The identification of the limits between each of these time units requires the individualization of stratigraphic layers, and thus of disruptions caused by global events; Ogg (2004) adds in his text "when possible". The challenge is to determine temporal intervals with significant changes in sedimentation, paleoenvironment and paleontological characteristics. It is, therefore, a matter of linking series of events of local or regional scale to a global scale (Cohen *et al.* 2013). However, the definition of the different units of the geologic time scale is done under the assumption that certain principles are met, including the principles of:

– continuity (equivalent age throughout the same stratigraphic layer defined by its lithological facies);

– actualism (past formation processes are equivalent to the present);

– paleontological identity (the discovery of identical fossils in two geological layers implies that they have the same age);

– superposition (apart from structural change, a layer is younger than the one it covers and older than the one that covers it);

– horizontality (the sedimentary layers are deposited horizontally; only a deformation subsequent to its deposition can disturb this position);

– intersection (the faults observed in the sedimentary layers are posterior to them);

– inclusion (the rocks observed in the sedimentary layers are prior to them).

The geological scale is thus made up of "single time surfaces", each with a precisely synchronous level that can be traced all around the Earth (Zalasiewicz *et al.* 2015). Yet, these principles are hardly ever respected, since (i) only more or less residual traces of these time units are traceable and (ii) most of the events constituting the different units are progressive, covering temporalities of thousands to millions of years (formation of mountain ranges, opening of the ocean, etc.) that include brief events (volcanic eruption, meteorite fall, etc.). Thus, in most cases, it is a matter of identifying the least poor range of indicators (Zalasiewicz *et al.* 2015) that leaves room for several arbitrary decisions in the absence of strict criteria, or the adoption of eclectic criteria (see Lucas (2018) for a history of the thinking leading to the constitution and revision of the geologic time scale). In the end, Lucas (2018) reminds us that the chronostratigraphy approach and method suffer from inconsistency, arbitrary decisions, reductionism, instability, opposition between global and local scales, imprecision and power struggles over the naming of different geological strata. The proposal of the Anthropocene as a chronostratigraphic or geochronological unit is an example of these power struggles. Indeed, the naming of geologic time scale units is subject to a vote by all members of the International Commission on Stratigraphy; however, the Anthropocene is not yet demonstrated by evidence of stratigraphic unity or boundaries recorded in lake or ice cores, or in other records at a global spatial scale (Finney and Edwards 2016).

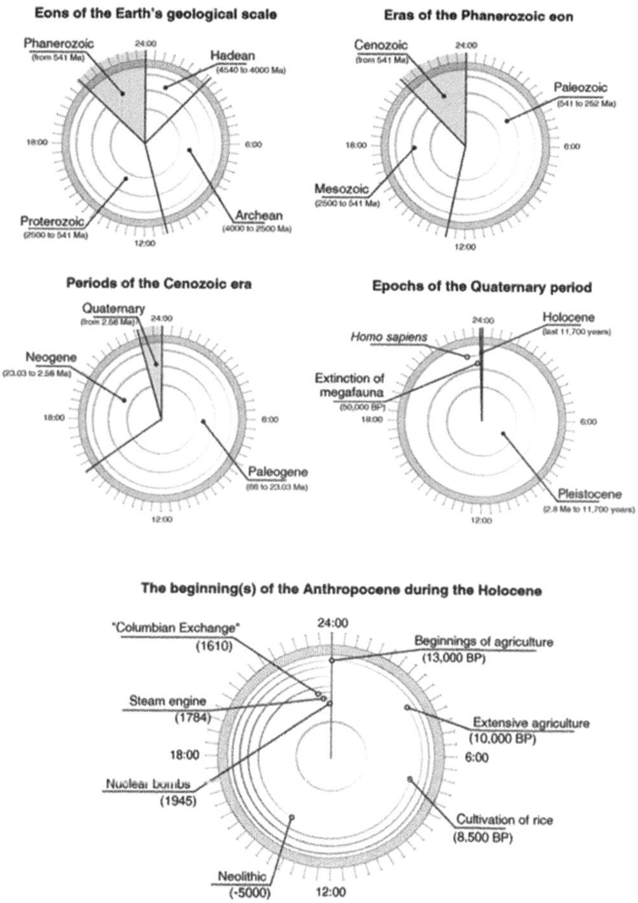

Figure 1.1. *The Anthropocene in the geological scale reduced to 24 hours, each of its subdivisions corresponding again to 24 hours (eons, eras, periods and epochs, then Anthropocene), showing the infinitesimal duration of the human impact on the environment during geological time, yet of considerable magnitude in its intensity*

1.4. When did we enter the Anthropocene?

While the stratigraphic approach struggles to propose a definitive answer to the existence of a particular geological period, other approaches have been mobilized to define a date of entry into the Anthropocene. The proposals for dates have been relatively numerous (Gemenne and Rankovic 2019) because the authors have sometimes emphasized the modification of one of the planet's components rather than another (very often the atmosphere and the biosphere, and also, to a lesser extent, the lithosphere or the hydrosphere). Also, they have been able to use different criteria to measure these changes (*e.g.*, for the atmosphere: the carbon dioxide, methane or nitrogen content; for the biosphere: the disappearance of species in one taxonomic group rather than another; for the lithosphere: the presence of radionuclides or micro-plastics, etc.). We present here, in chronological order, the main dates used in the literature to identify the beginning of the Anthropocene (see Figure 1.1).

1.4.1. *50,000 years BP: the end of the Pleistocene and the extinction of megafauna*

Extinction crises, generally understood as the global disappearance of three quarters or more of the species in a short period of time, are strong indicators of major abrupt changes on a planetary scale. They are, in themselves, moments of change in the Earth's system because a large part of the biosphere is profoundly modified. The extinction crises of the living world are also revealing because they represent one of the consequences of changes in a set of components of the planet. The five extinction crises that followed one another during the history of the Earth are indeed marked by important climatic changes, sometimes sudden marine transgressions and regressions, tectonic and volcanic phenomena, asteroid impacts or massive anoxia of water masses; all these modifications are intimately linked to one another (see Barnosky *et al.* (2011)).

The beginning of the sixth extinction crisis that we are currently experiencing can be located at the end of the Pleistocene with the extinction of megafauna (see Chapter 5 of this book). It was during this period that animals weighing more than 44 kg as adults (see Martin (1984)) disappeared in fairly short periods of time. The origin of these disappearances is still debated. Several camps are in conflict, with, schematically, the proponents of an anthropic origin (Martin 1967, 1984), those of an environmental origin (see, for example, Grayson and Meltzer (2003) concerning the case of North America) or those favoring a combination of the two (see, for example, Elias and Schreve (2007)). The debate is likely to remain sterile in the absence of dating and close examination of new fossils.

The end of the Pleistocene can thus be seen as the beginning of the Anthropocene if we consider that it is indeed the beginning of the sixth extinction crisis of living organisms and that its origin is essentially anthropogenic. In addition to the question of the anthropogenic character of the crisis, the identification of this period as the starting point of the Anthropocene poses a problem because the sixth crisis is currently underway and even seems to be accelerating and affecting an ever larger number of taxonomic groups. It is therefore difficult to identify the limits of a crisis of which we do not yet know everything and whose climax is perhaps still to come. The speed of this crisis is, however, indisputable. While the previous biodiversity crises were generally spread over several million years (with the exception of the Cretaceous crisis, during which the impact of the asteroid in the Yucatan probably had rapid effects), the sixth crisis is already identifiable in only a few thousand years.

1.4.2. *5–7000 years BP: the Neolithic and the increase of methane and CO_2*

While hunter-gatherers were most likely responsible for, or at least contributed to, the disappearance of the Pleistocene megafauna, the changes made during the Neolithic period were even more significant. This period marked a change in the way of inhabiting and exploiting the Earth, and thus in the intensity and extent of the modifications brought by humans to the planet. The sedentarization of human populations, the clearing of land, the development of agriculture, associated with the storage of harvested products and the beginning of animal husbandry were changes that took place in several parts of the world and at dates beginning at the earliest around 11,000 years BP in the Near East or 6500 years BP in Europe.

The Early Anthropogenic Hypothesis (EAH) was first published nearly 20 years ago (Ruddiman 2003), stating that the influence of humans on climate began very early, with the increase in methane concentrations 5,000 years ago due to anthropogenic emissions from large-scale rice cultivation and cattle ranching, and the increase in atmospheric CO_2 that began 7,000 years ago due to deforestation.

The anthropogenic origin of the methane increase has been fairly uncontradicted (but still see the orbital forcing hypothesis by Singarayer *et al.* (2011)). Evidence was published by Fuller *et al.* (2011), who mapped the development of rice cultivation in Asia between 5,000 and 1,000 years ago. They estimated that CH_4 emissions from northern rice fields were responsible for 70% of the increase in atmospheric CH_4 recorded in ice cores during this time interval. Today, the IPCC is also adopting the Ruddiman hypothesis (see Ciais *et al.* (2013)).

In contrast, the hypothesis of increased CO_2 7,000 years ago due to humans has been more controversial. Ruddiman *et al.* (2020) defend this hypothesis on the basis of demographic data. Early demographic data documented a global human population of the order of a few million people 10,000 years ago (McEvedy and Jones (1978) cited in Ruddiman *et al.* (2020)), followed by a doubling of this population every thousand years, bringing the world population to over a billion people a few centuries ago. According to this progression, the planetary human population was relatively small 7,000 years ago and would therefore not have started its spectacular progression. However, recent archaeological studies conducted in China have shown a completely different population dynamic (see, for example, Li *et al.* (2008); Hosner *et al.* (2016)) with a marked demographic increase as early as 7,000 years BP, at a time when the hunter-gatherers of this region of the world became farmers, and began to practice deforestation in order to establish crops and pastures. In Europe, numerous archeological sites have been uncovered, corresponding to dates between 5,000 and 7,000 years BP, coinciding with the first settlements of farmers and herders (Ruddiman *et al.* 2020). This increase in population in Europe and China, two major population centers at that time, accompanied by significant deforestation, lends credence to the increase in anthropogenic atmospheric CO_2 during the same period.

However, the amount of CO_2 emissions has been much debated. Some authors (see Table 1 in Ruddiman *et al.* (2020) for a list of these studies) consider Ruddiman's (2003) estimates to be overestimated by a factor of 5. It is clear that extensive ancient deforestation occurred at several points around the globe, including after 6,000 years BP in Europe (Fyfe *et al.* 2014; Roberts *et al.* 2018). Deforested areas remain difficult to assess because they are based on average areas estimated to be needed by a farmer, and what is meant by "deforested area" is also subject to controversy (*e.g.*, either only areas converted to crops and pasture, or partially or temporarily deforested areas – see Houghton and Hackler (2003)). In contrast, paleoenvironmental reconstructions provide evidence for a significant transformation of valley floor morphology in response to deforestation in many European rivers (Lespez 2012; Brown *et al.* 2018).

1.4.3. *1610: "Columbian exchange", low CO_2 level and cooling of the Little Ice Age*

It is the year 1610 that is promoted by Lewis and Maslin (2015, 2018) as the official start date of the Anthropocene. According to these authors, the 17th century was marked by visible evidence of the connection of previously separated continents and oceans, and thus an exchange of species across the world. This "Columbian exchange" (a term coined by Crosby (2003) to describe the exchanges between Old and New Continents that occurred following the "discovery" of America by

Christopher Columbus in 1492) is well identified in stratigraphic observations and archaeological materials. However, Lewis and Maslin (2018) prefer to follow the prevailing recommendations of the geological community, and thus do not use a biostratigraphic marker as the GSSP of the Anthropocene. They therefore advocate a GSSP based on a change in atmospheric composition, specifically the tipping point in the amount of atmospheric CO_2 due to the arrival of Europeans in America. Following their arrival, the Europeans unwittingly decimated the native populations mainly by importing the smallpox virus, which led to the abandonment of agricultural land and the development of a vegetation cover that stored a significant amount of carbon, while causing a drop in carbon in the atmosphere. This drop has in fact been well recorded in ice cores, particularly from Antarctica (Dull *et al.* 2010). Lewis and Maslin (2015, 2018), therefore, use the minimum CO_2 values from this period recorded in the Law Dome ice core, 1610, to give an Anthropocene start date. This corresponds to the coldest period of the Little Ice Age. The year 1610 is, therefore, a tipping point where the fauna and flora of several continents that had been separated until then mixed, and also where the CO_2 reached a very low level and temperatures were particularly low.

1.4.4. *End of the 18th century: the First Industrial Revolution*

One of the most commonly cited dates for the beginning of the Anthropocene corresponds to the end of the 18th century and the beginning of the 19th century (see Bonneuil and Fressoz (2013)). This is the period proposed by Crutzen, to whom the name of this new phase of the Earth's geological history belongs. For Crutzen (2002), it is indeed from the end of the 18th century that we note the beginning of an increase in the atmospheric concentration of CO_2 and methane. This corresponds to the patenting of the steam engine by James Watt in 1784 and the entry into the First Industrial Revolution. In Europe, this period marked the transition from slow population growth, an agricultural society, and the use of energy derived primarily from human muscle power and wood burning to rapid population growth, an urban and industry-oriented society processing fossil fuels (Zalasiewicz *et al.* 2015).

1.4.5. *The mid-20th century: the great acceleration and the fallout of radionuclides*

In the aftermath of World War II, the mid-20th century was marked by a major population boom, economic growth and environmental changes (Zalasiewicz *et al.* 2015). This period is generally referred to as the "Great Acceleration" (see Steffen *et al.* (2007)). It is characterized by several aspects: a major increase in CO_2 emissions since the pre-industrial era (higher levels of CO_2 were recorded in the Pliocene, four million years earlier), the democratization of the automobile, agricultural intensification with significant use of chemical fertilizers and a globalization phenomenon (Zalasiewicz *et al.* 2015). The environmental change generated by these changes far beyond just the inhabited urban centers of humans has been the basis for identifying a Holocene/Anthropocene divide (see Wolfe *et al.* (2013)). McNeill and Engelke (2016) wrote an entire book entitled *The Great Acceleration: An Environmental History of the Anthropocene Since 1945*. In that book, despite the consensus of the illusion of determining a precise date at the beginning of the Anthropocene, the authors nevertheless propose to consider the middle of the 20th century (around 1945–1950) as the beginning of this period. Although they point out that humans have always modified their environment, the authors successively review: energy and human population, climate and biodiversity, cities and the economy, the Cold War and environmental culture, in order to demonstrate that the influence of humans took on a whole new dimension from 1945 to 1950. It is from this period that the concentration of CO_2 reached levels never reached in 870,000 years, the nitrogen cycle was disrupted (see Canfield *et al.* (2010)[1]), and the anthropogenic impact on the biosphere took on a major magnitude, notably through the unprecedented growth of cities or the massive construction of dams, the acidification of the oceans or the accumulation of plastic debris. They point out that if the Great Acceleration on the one hand and the Anthropocene on the other coincide in time, it will not last. The Great Acceleration will come to an end: the human population cannot continue to grow and the use of fossil resources is almost at an end. On the contrary, the Anthropocene will continue longer, because for McNeill and Engelke (2016): "Indeed, even if every human immigrated to another planet tomorrow, our impacts of the past few generations will linger for millennia in the Earth's crust, in the fossil record, and in climate."

Finally, to support the idea of the beginning of the Anthropocene around the 1950s, let us point out that the votes of the working group on the Anthropocene of the International Commission on Stratigraphy made known on May 21, 2019 that the basis of the Anthropocene lies on the stratigraphic signals of the mid-20th century.

The decade that followed, the 1960s, has also been proposed as the starting point of the Anthropocene. Specifically, the year 1964, the year of peak radionuclide fallout, has been proposed as the GSSP by several authors (Lewis and Maslin 2015, 2018; Waters *et al.* 2016) and referred to as Bomb Spike (Lewis and Maslin 2018). For Lewis and Maslin (2018), this radionuclide fallout is recorded in many repositories all over the Earth and will likely be present for millions of years. Lewis

1 The invention of the Haber-Bosch process in the early 20th century, enabling the conversion of atmospheric nitrogen to ammonia for soil fertilization, has altered the global nitrogen cycle so much that the closest geological comparison in modern times that can be suggested refers to events that occurred about 2.5 billion years ago.

and Maslin (2018) believe that they coincide well with the Great Acceleration, and thus they are somewhat at odds with Zalasiewicz *et al.* (2015) or McNeill and Engelke (2016) for whom, as noted, the Great Acceleration began instead in the 1940s–1950s.

1.5. A temporal baseline on the fringe of the Anthropocene

By definition, the Anthropocene concept focuses on the question of the anthropization of environments. The tipping point notion leads to the search for an absolute baseline in the past, *i.e.*, a state free of any anthropic influence or with a minimal level of transformation. Nevertheless, another use of the notion of baseline exists, without necessarily considering its natural character.

Indeed, a past biophysical functioning can be considered as a baseline if it corresponds to a socially desired functioning for the services it provides or for its relative stability. Thus, the work of the Mountain Land Restoration services in France (*Restauration des Terrains en Montagne*, RTM) was implemented as early as the 19th century to reduce the risks associated with sediment production (river bed embankment and associated flood risks, slope instability, etc.). It is a question of "restoring to the cleared land its primitive adhesion" (Dugied (1818) cited in Combes (1989); Hall (2005)). RTM mobilizes a state prior to agricultural clearings that is considered more socially acceptable because it has the advantage of reducing exposure to hazards (through a management perspective that is very close to engineering); the acceptance of RTM interventions is therefore not linked to the prospect of a return to a certain naturalness (Brugnot 2002).

A baseline can also correspond to a landscape, a functioning, characteristics that are distinguished by their rarity and their intensity. For example, in the French or Italian Alps, braided rivers are currently the subject of much attention because they generate emblematic landscapes, present intense biomorphological processes and shelter particular plant and animal species. However, their establishment is partly based on anthropogenic modifications of the mountain slopes during the 19th century (Gautier 1994; Roche *et al.* 2005; Liébault *et al.* 2010). Today, this fluvial style is becoming scarcer, in part due to reduced sediment production. Among the causes of this sediment deficit, spontaneous reforestation following the decrease in anthropogenic pressures (*e.g.*, cessation of grazing) is most often perceived or implied as a limit to the possibility of restoring a river to its most intense morphological functioning, and thus "idealized", or even as an alteration of a process rather than a return to more naturalness (*e.g.*, Surian and Rinaldi 2004; Brousse *et al.* 2019). The baseline of these systems was often sought in the 19th century, a period in which direct anthropic pressure was nevertheless the most significant (Hall 2005).

1.6. Conclusion

To define the Anthropocene is to admit that humans mark a period in the history of the Earth. This is not arrogance, and this is for two reasons. First, it is about recognizing the facts (changes in the atmosphere, the lithosphere, biodiversity, etc.); but it is also about recognizing the negative consequences of our species' actions on all the other compartments of the Earth system, as well as on our own species. The only arrogance is probably to "put all humans in the same boat" of the Anthropocene: they do have a common destiny, but only a fraction of humanity is at the origin of it.

The first difficulty in defining the starting point of the Anthropocene is related to the extremely recent character of all the events we have been able to list (from the end of the Pleistocene to the 1960s). The data concerning the last few thousand years are much more diverse and abundant than the older ones, which can lead to a much finer identification of recent events compared to the older ones. The second major difficulty in identifying the Anthropocene is that it is the contemporary period. How can we then know if the greatest changes are not yet to come? How can we analyze and conclude the identification of a period that is not yet over?

The question of the irreversibility of harmful, anthropogenic actions also arises. The lockdown of spring 2020 due to the Covid-19 pandemic and the almost total and abrupt ceasing of human activities on a global scale, at least concerning the western world, has shown that the impact of humans is not necessarily permanent on certain aspects (reduction of the pollution of the Venice lagoon; return of the observation of great fauna; decrease in urban smog, etc.). What is the temporality of the chronostratigraphic limits between the different geological layers? What is the duration of the "sudden changes" justifying the passage from one layer to another?

In the end, the Earth is 4.54 billion years old; *Homo sapiens* appeared 300,000 years ago; they are therefore present on only 0.0066% of the geological scale and have undoubtedly profoundly modified the Earth system. If we consider the low limit of the beginning of the Anthropocene (around 50,000 years ago with the extinction of the megafauna), it represents 0.00011% of the geological scale. Considering its high limit (around 1950, or 70 years), it is 0.0000015%. Brought back to a scale of 24 hours, these values represent 5.7–0.09 seconds, *i.e.*, for the low value, hardly the duration of a blink of the eye.

2

Spatial Baselines: Is Going Elsewhere Easier Than Going Back in Time?

2.1. Preamble

We completed our PhD in the early 2000s, and during this time, we attended several conferences dedicated to river management, ecological restoration of wetlands, the historical geography of forests, etc. Many of the discussions held at these events were related to the baseline question, either during the presentation sessions or during more informal moments. What was the content of the exchanges? Without reporting on all of them, we will note a common feature of most of these discussions: they addressed the question of how the objectives of a management plan or restoration action are defined. And almost systematically a debate followed: "for us it is quite simple, because we have data on the state of our wetland site before it was transformed by human activities", "for us it is more difficult, since our site has been occupied by humans for several centuries, so we have surveyed other comparable, but more natural, sites", "we rely on models that predict the type of environmental conditions and biological communities that should be present without anthropogenic pressure", etc. It seems to us that there was some confusion due to different visions of what is "natural", the history of sites, what is a comparable site, etc. This chapter aims to propose elements of clarification that we would have liked to have had two decades ago in order to facilitate exchanges.

2.2. Introduction

The baseline has often been defined as a state without disturbance, *i.e.*, without pressure from human activities (*e.g.*, NRC 1992). Therefore, the first impulse was to

Chapter written by Anne-Julia ROLLET, Simon DUFOUR and Armelle DECAULNE.

look for these baselines in the past, before anthropic pressure profoundly transformed the bio-physical functioning of the environments. If this reference to the past fits perfectly with the old adage "it was better before", the quest for what the English-language bibliography describes as a *pristine* state quickly encountered numerous obstacles. These obstacles appear both in the construction of the baseline and in its application. Regarding the first point, it is, for instance, difficult to obtain historical data, which are scarce and often imprecise (Schletterrer *et al.* 2014), and to temporally define the baseline period (how far back in time?). The second point concerns the relevance of the concrete use of these historical baselines as a management tool in contexts where anthropic pressures are multiple, old, and when the dynamics of environmental conditions no longer enable the realization of this baseline. For example, Wyzga *et al.* (2012) were able to observe, in Polish Carpathian streams, significant reductions in flows and sediment inputs related to environmental changes within watersheds characterized by very little direct anthropogenic influence. Even non-anthropogenic ecosystems are dynamic! Thus, while the historical approach provides relevant information on the trajectory of a site, restoring a system to its pre-disturbance conditions is, strictly speaking, impossible in most situations.

To get around these limitations, another approach is to look for baseline systems elsewhere in space, in areas that are still relatively preserved, or where there is a consensus on the quality or integrity of current biophysical functioning. However, behind this apparent simplicity, the spatial dimension is not simpler or smoother than the temporal dimension, and this raises many eminently geographical questions such as the representativeness of the selected sites, the spatial scale at which these baselines should be sought (sites? eco-regions?) or the modalities for integrating, or not, the anthropic influence on the selected sites.

In this chapter, we propose to first present these approaches to the notion of spatial baseline, and in particular, the criteria used for the selection of such sites. Second, we will illustrate these approaches with detailed examples of definitions of single or multiple baseline sites. Finally, the limits and interests of these approaches will be discussed[1].

1 This chapter aims to be relatively universal in its definition and use of baseline sites through a spatial approach. Nevertheless, the examples will often concern rivers, given the importance of the issues of scale, levels of anthropic influence and site representativeness in the scientific work developed on these environments. Moreover, the question of spatial baseline has been particularly and explicitly addressed in the context of the restoration of these environments.

2.3. What is a spatial baseline?

The notion of spatial baseline is based on a (seemingly) simple principle: to identify, elsewhere, a site (or sites) with characteristics considered desirable. These characteristics can then be used either to establish the degradation level of the site studied, by comparison, or to define management or restoration objectives. However, the selection of these baseline sites is a complex process that can be implemented in many ways and is not easily freed from cultural and emotional characteristics; this was indeed identified by George Perkins Marsh as early as 1856; this pioneer of conservationist thinking and restoration then remarked that the regions of southern Europe lack "any thing which corresponds with an American's idea of a forest" (Hall 2005, p. 18).

2.3.1. *In search of naturalness...*

It seems that the first criterion of importance for authors seeking baseline sites through a spatial approach is that of naturalness: the less a site has been subjected to pressure from human activities, the more the ecological conditions encountered there will be considered as a baseline. In the vast majority of studies, the justification for site selection is based on an argument relating to the low level of pressure and/or transformation and, in some cases (*e.g.*, Herlihy *et al.* 2008; Shinneman *et al.* 2008), this is even the only criterion explicitly presented. There are, however, several ways to consider this naturalness. For example, the baseline conditions correspond to the attributes that a site would have if it had not been transformed by human activities, transformations being considered as degradations. In this logic, a site where pressure is non-existent corresponds to a good state of conservation, good integrity, and therefore necessarily, to good functioning (Armanini *et al.* 2013). Conversely, the deviation from "good status" would necessarily be induced by an anthropic influence (Roche *et al.* 2005). While the approach seems relatively effective, several difficulties (or biases) are regularly mentioned in the literature. The main difficulty lies in the idea that there are no longer any natural environments in the sense of intact environments, devoid of anthropic influences, an idea that seems to be widely recognized in the scientific community today: systems profoundly transformed by the arrival of new species (Schletterer *et al.* 2014), legacy effects via very old influences whose effects are still perceptible (Dupouey *et al.* 2002), "remote" effects linked to pollution phenomena or climate change, etc. However, a certain ambiguity remains in the discourses or in the minds. For instance, the Tagliamento River (Italy) is often considered as a reference river for the Alps, especially because of its "natural" character (see Figure 2.1). It is thus referred to as "Europe's last great natural Alpine river" (Müller 1995) or "the only large morphologically intact Alpine river remaining in Europe [...] provides insight into the natural dynamics and complexity that must have characterized Alpine rivers in the pristine state". (Ward *et al.* 1999). However, a careful review of these papers shows that the same authors use significantly different terms in their subsequent discussions, such as near natural river segments, semi-natural reaches, relatively pristine sites, or "a river [that] is not entirely without human impact" (Ward *et al.* 1999). It thus seems that this quasi-theoretical baseline system is not so pure, without fully assuming the integration of the anthropization of its watershed.

Figure 2.1. *The Tagliamento River celebrated for its wildness in a WWF document (date unknown; source: https://wilderness-society.org/wild-river-tagliamento-in-northern-italy). For a color version of this figure, see www.iste.co.uk/godet/baselines.zip*

Since sites without anthropogenic influence are now almost impossible to find, researchers turn to the least affected sites and compose a baseline from the best possible conditions within a given area (Herlihy *et al.* 2008). This is particularly common when establishing baselines in fine-scale contexts or contexts with particularly strong agricultural or urban identities, for which the "least disturbed" sites are selected (*e.g.*, Kershner *et al.* 2004; Down *et al.* 2011). The authors distinguish between minimally and least disturbed sites (Whittier *et al.* 2007) or baseline and default baseline sites (Mengin *et al.* 2009). The aim is, therefore, to select baseline sites where the biocenoses show different levels of alteration but remain relatively functional, at least on certain criteria, and no longer sites where no anthropogenic influence is observed (Raven *et al.* 2002; Whittier *et al.* 2007; Armanini *et al.* 2013). The goal is no longer to establish a "perfect" baseline, but to model what could be achieved "at best" in contexts where anthropogenic pressure at the watershed scale is considered non-suppressible.

However, in other contexts, the anthropogenic pressure is so great that it is difficult to find areas that can become a baseline, even in a mitigated or relative sense. For instance, in their search for baseline river sections in Denmark, Baattrup-Pedersen *et al.* (2009) observed that almost no site could meet the classically predefined requirements for low anthropogenic pressure, especially because of criteria that are particularly disqualifying at the watershed level: land artificialization, intensive agriculture, nitrate concentration etc. And when some sites manage to meet these criteria, the selected sites often constitute only partial baselines insofar as they are only qualified for certain taxonomic groups (*e.g.*, macrophytes and macro-invertebrates), and cannot constitute global baselines.

2.3.2. *On the basis of which indicators?*

Most often, anthropogenic pressure is defined on the basis of criteria that characterize direct, visible, recent impacts (Wallin *et al.* 2003; Shinneman *et al.* 2008), for instance, through information on land use, the existence of pollution or certain practices (*e.g.*, logging, drainage, grazing), or the presence of infrastructure (dams, roads). However, some authors propose not to rely on the presence/absence of visible anthropogenic elements to presume the quality of an environment, and thus recommend not to systematically exclude sites where anthropogenic pressure is observed (*e.g.*, Wallin *et al.* 2003). They propose first to evaluate the real effects of this pressure on the biophysical characteristics of a site and tolerate the pressure as long as its ecological consequences remain null or very weak, *i.e.*, as long as it remains difficult to discern them from the natural variability of the parameters studied (Mengin *et al.* 2009). The selection criteria then shift from pressure to the response of ecological systems and the approach mobilizes statistical models that scientifically establish a link between anthropogenic pressure and the functional characteristics of a site.

Whatever approach is chosen (pressure or response analysis), it requires a very explicit definition of the values/criteria that will make it possible to select a site as a baseline or not. There are national and international inter-calibration approaches that make it possible to define "threshold values" for different standard criteria that are homogenized in this way. For example, we can cite the European grid of biological indicators for rivers developed by Pardo *et al.* (2012), which makes the methods used comparable between member states; the regional grid for riparian forests in the NW of the United States of America (Pollock *et al.* 2012); or the common protocol of the ITEX (International Tundra Experiment) polar observation network (Molau and Molgaard 1996).

In terms of data sources for the production of indicators, the quest for desired conditions mainly involves two approaches: selection by expert opinion and selection by quantified criteria (Huggins and Dzialowski 2005). The "expert judgment" approach can be criticized for including sites that are not truly benchmarks because of the overemphasis on one parameter over all others, based on the potential preconceptions or concerns of the experts involved (Whittier *et al.* 2007). For instance, Mengin *et al.* (2009) suggest that the focus should be on the functionality of sites rather than their richness, since a functional site may host a limited number of taxa. For several years, this expert approach seems to have regressed in favor of selection based on quantified criteria fed by metrics measured directly in the field and/or by landscape analysis and the production of metrics by GIS. The increased use of quantitative approaches relies largely on the development of spatial analysis tools and the resolution of remote sensing data. Far from being mutually exclusive, the complementary use of these metrics (field measurements and geomatics) seems relevant in many cases (Lunde *et al.* 2013). It is even sometimes complemented by local expertise (local residents, site managers; Guyon and Battaglia 2018). Quantified criteria approaches are, indeed, not free of limitations that impact the adequacy of the baseline: (i) inadequate or inaccurate GIS layers, (ii) limited or inaccurate information on range-wide stressors and (iii) inadequate or uneven sampling effort (Ode *et al.* 2014).

2.3.3. *In search of truly comparable sites*

While the question of the naturalness degree is common to both spatial and temporal approaches of the baseline notion, there is another issue that is more specific to spatial approaches. It is a question of ensuring that the baseline conditions come from a context that is comparable to the site being studied, what Soar and Thorne (2001) call a "carbon copy". For instance, for river systems, this means looking for baseline sites with similar characteristics on a nearby river reach, either upstream or downstream (see Rosgen (1994)), or on distant river reaches with similar watershed size, lithology, slope, etc. (Montgomery *et al.* 1995; Montgomery and MacDonald 2002).

The search for baseline sites can be done either on a case-by-case basis, or based on classifications that facilitate the identification of comparable conditions. For this purpose, the sites considered as not degraded belong either to the same type or to the same ecoregion as the site studied:

– The type approach is a form of classification describing categories of ecological, landscape-related, or topographic and geomorphic units distinguished on the basis of their intrinsic properties, but not explicitly their location, such as a habitat (*e.g.*, Oligotrophic grasslands) (Clair *et al.* 2019), fluvial style (*e.g.*, braided reaches) (Brierley and Fryirs 2005) or snow-avalanche behavior (*e.g.*, McClung and Lied 1987; McClung and Mears 1995) (see Figure 2.2A).

The region approach is a form of explicitly spatial classification that generally describes large, single geographic units; these are referred to as ecoregions, hydroecoregions, etc. (Omernik 1987; Wasson *et al.* 2002; Hawkin *et al.* 2010), within which sites are considered comparable because they are subject to the same general conditions (climate, geologic substrate, geomorphic functioning, etc.) (see Figure 2.2B). However, the regional scale encompasses great disparities and such an approach neutralizes site effects, which are essential for biotic and abiotic behavior, since environmental conditions can change drastically over short distances. For example, the use of lichenometry as a means of dating recent geomorphic landforms in cold high-latitude environments has shown strong limitations related to environmental variability over very small distances (a few meters to a few hundred meters), which has consequences for the growth and mortality of individuals (McCarthy 2013; Osborn *et al.* 2015); this has been documented in detail in Iceland (Decaulne 2016), where obtaining a local baseline is advocated proposing absolute dating over a period of only a few decades, with dating being relative beyond a century.

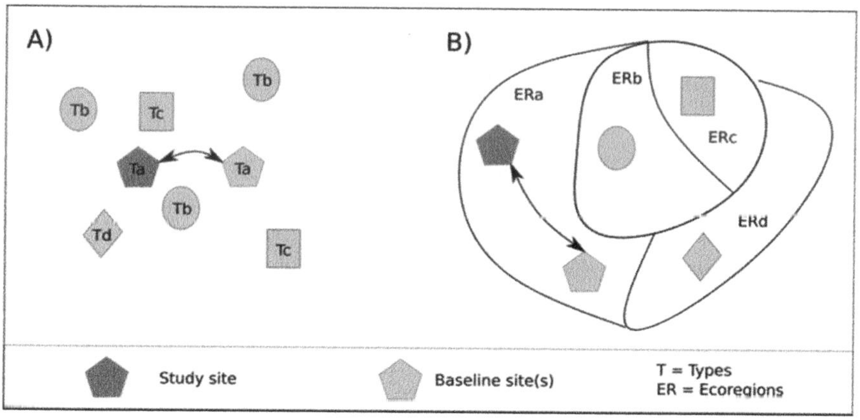

Figure 2.2. *Identification of similar conditions at the study site: A) by typological approach and B) by regionalized approach*

However, finding comparable contexts is not easy. On the contrary, White and Walker (1997) note that in application of the first law of geography, it is theoretically impossible to find perfectly comparable conditions because we must move away from the degraded site and, in so doing, the probability of encountering a strictly similar site decreases. We cannot find two sites that are identical in every respect: one in good condition and the other in poor condition! Hence, it is common to use several sites in order to not establish a baseline, but a range of baseline conditions (see below). On the contrary, the sites considered to be of good quality are never homogeneously distributed. For instance, in the case of the semi-arid landscapes of western Colorado (USA), Shinneman *et al.* (2008) found baseline sites mainly in protected areas, whose spatial distribution over this vast territory does not enable the representation of all the conditions of the region being studied. Similarly, on French metropolitan rivers, since pressures are frequently lower upstream of the watersheds, the distribution of baseline sites leads to an over-representation of medium and small rivers compared to larger rivers, which remain chronically less referenced (Wallin *et al.* 2003; Mengin *et al.* 2009). Lowland streams are also often overlooked because they are known to be historically and heavily anthropized (flow regulation, agricultural developments, pollution, etc.; Petts *et al.* 1989).

Ultimately, if the site selection criteria are defined too restrictively, they are too limiting and the number of eligible reaches is too small to ensure statistical validity of the baseline. Ode *et al.* (2014) give the example of California streams where, if the low pressure criteria are defined in a demanding way, only sites in mountainous regions (Sierra Nevada and North Coast Mountains) are selected, yet they are not representative of most California streams. However, these criteria must be sufficiently stringent to be meaningful for establishing environmental quality. It is, therefore, necessary to find a compromise between demanding pressure intensity criteria and the minimum number of sites selected.

2.3.4. *A single site or a collection of sites?*

In addition to the definition of the essence of the baseline (*e.g.*, a system functioning free of influences, the least degraded possible), the type of relevant indicators and the similarity between study and baseline contexts, it is finally necessary to choose between two possible options: does the baseline come from a single site, which alone represents a baseline, or from a collection of baseline sites[2]?

2 This distinction remains schematic and debatable because the notion of site varies and, above all, does not correspond to a homogeneous spatial coverage according to the authors. Thus, a forest may be considered as a single site by one author, but be made up of several plots or ecological units. In this chapter, the "single-site" character corresponds to cases where the authors consider a territory as providing a baseline for a category considered as homogeneous a priori. The "multiple site" character will be reserved for approaches that explicitly select several sites, usually remote from each other, in order to cover a posteriori the maximum diversity of possible conditions.

In the case where only one site is used as a baseline, it is necessary to distinguish the case of the "twin" site from the case of the "showcase" site (see Figure 2.3):

– The first one corresponds to a non- or slightly degraded site selected according to either its geographical proximity or the proximity of its environmental characteristics to those of the site studied. The biophysical characteristics of this twin site are then directly used as a baseline (topography, geomorphological processes, species list, size and/or age of trees, etc.). Dendrochronological investigations carried out, for instance, in sites affected by snow-avalanche dynamics, or by insect attack, use the growth pattern of tree-rings not exposed to the disturbing growth process, *i.e.*, those showing baseline growth, to individualize the ring growths showing deformation due to the disruptive process. This eliminates environmental factors (precipitation, temperature, soil moisture, etc.) common to all trees (affected or not by the disruptive processes) in the study area (Filion and Gärtner 2010; Speers 2011). This twin approach, which is quite intuitive, seems to be in the minority in the scientific literature, and limited to only certain research areas (see Hawkins *et al.* (2010)); it results in too few data to establish robust[3] statistical models. Moreover, the baseline thus defined is rarely transferable/transposable to other remote sites, as it requires an undeniable proximity character.

– The notion of a "showcase" site is based on the identification of a site considered exceptional, often archetypal, which would represent the absolute baseline for a type of site or environment: the steppe formations of the Crau plain, the wetlands of the Camargue, the Białowieża Forest, etc. It is then interesting to note that the site is attributed a (quasi-)universal scope, while it is most often of an extraordinary, "out of the ordinary" character. Showcase sites are also used as a basis for building and validating conceptual models (structural and/or functional): these sites are used as full-scale laboratories to build, test and validate scientific models that are then used as a baseline (see section 2.4). One of the major difficulties of this approach is the lack of scientific consensus on the sites selected, due, in particular, to the diversity of disciplines and research fields concerned in the same area. The use of a single site as a baseline also raises the question of the validity of the comparison, and therefore of the integration of the (natural) diversity of environments, which do not enable the definition of unique baseline conditions for all sites within vast territories (regional, national, etc.). Finally, the construction of the spatial baseline from a single site, twin or showcase, is very sensitive to the uniqueness of trajectories (SER 2004). Thus, the value of a single site is necessarily spatially limited; it is problematic to consider it as a baseline that is easily transposable to different contexts. In response to these limitations, other approaches, based on multiple sites, have been developed.

---

3 Bowman and Somers (2006) recommend a minimum of 20 sites, but preferably 30–50, for establishing baseline conditions in a given setting.

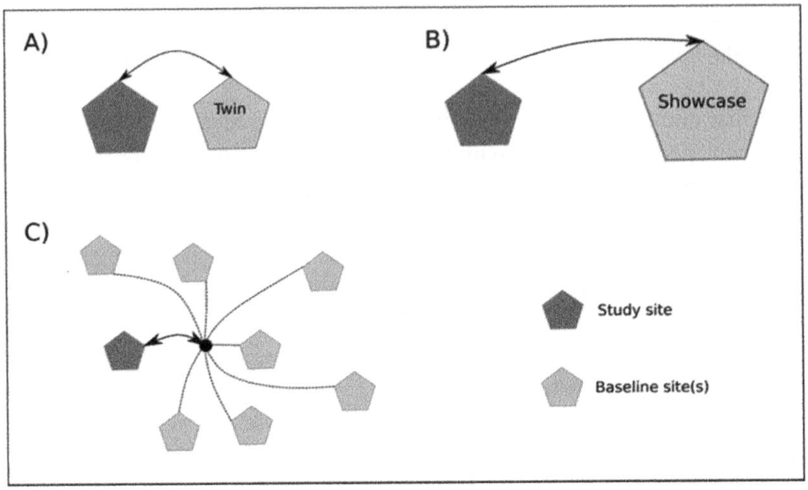

Figure 2.3. *Different approaches to selecting a baseline site(s) for a study site: A) single-site approach with a sister site, B) single-site approach with a showcase site and C) multi-site approach*

Multi-site approaches involve selecting a collection of baseline sites from which average values and ranges of variability can be derived for parameters of interest (see Figure 2.3C). These ranges of values are then used as a baseline to either establish a target or assess a condition by comparison (*e.g.*, Rheinhardt 2007). Defining baseline conditions on the basis of ranges of variability, rather than single values, offers the possibility of capturing the natural variability of measured parameters. Note that this consideration of variability can be done through a spatial input (by multiplying the baseline sites) and also temporal (by repeating the measurements of community/process metrics within a smaller number of sites). For example, Schletterer *et al.* (2014) constructed a spatial baseline on the Volga River from iterating metric measurements at selected baseline sites over a five-year period to incorporate intra- and inter-annual variability in zoobenthos. This repetition of measurements over time is a classic approach in climatology with the notion of temperature anomaly. It is relevant for establishing a diagnosis of the potential of a site to constitute a baseline, especially in contexts of high temporal variability (forest fires, intermittent watercourses) (Dallas 2013). In the study of slope dynamics, and more specifically avalanches and their travel lengths, it is the accumulation of information from different paths that determines the baseline (*e.g.*, Lied and Bakkehøi 1980; Bakkehøi *et al.* 1983; McClung *et al.* 1989; Delparte 2008): regionally, the analysis of topographic parameters on multiple slopes documenting multiple snow-avalanche paths defines a baseline snow-avalanche behavior, transferable to other slopes. Either the latter have never been documented in the archives (no known snow avalanche in such a path), or the distal zone has been modified by humans before the knowledge of previous snow-avalanche events, and the topographic parameters of the bottom of the slope are truncated. The use of a baseline defined by a regional multi-site approach is then relevant.

2.4. Emblematic examples of single- and multi-site spatial baselines

2.4.1. *The Białowieża Forest, a baseline for European forests?*

> TV5Monde: Why is this type of forest important?
>
> Francis Hallé: [primary forests have] ecological importance. In terms of carbon storage, you can't get much better than a primary forest. Another aspect that is very close to my heart is that as the forest ages – and therefore approaches the primary state – biodiversity increases. The primary forest is the pinnacle of biological diversity in any part of the world, whether in temperate or tropical countries. There is also an aesthetic aspect to which I am very sensitive. The forests we have in Europe are not beautiful: they are secondary forests. We have had them in front of us for so long that we think they are beautiful, but that is because we do not have the baseline in this matter, which is the primary forest. There is one that you should rush to see for those who are interested, and that is the Białowieża Forest in Poland. It is absolutely beautiful and unforgettable.[4]

Hallé's words testify to the place that the Białowieża Forest holds for scientists and activists in the field of forest conservation and management. Why is there such interest in this forest? The vocabulary used to describe Białowieża[5] is illuminating: not only is it considered virtually free of direct anthropogenic influences ("primary forest", "quasi-primary forest", "not degraded by human activities", "pristine forest", "[…] has survived in its natural state until today" according to the sources), but above all it is presented as the only surviving example of Central European lowland mixed forest in this state: "a unique degree of naturalness", "one of the most natural forests in Europe", "a unique forest because it originates from natural dynamics without substantial anthropogenic influence", "an exceptional importance

4 Translation of an interview given by the French biologist, Francis Hallé, to TV5Monde for his association "L'association Francis Hallé pour la forêt primaire". Available at: https://information.tv5monde.com/info/ecologie-une-foret-primaire-de-70-000-hectares-pourrait-elle-voir-le-jour-en-europe-de-l-ouest.

5 At least a part of the forest, because the Białowieża Forest massif consists of several areas under protection status and with varying degrees of direct human intervention.

for conservation", "an irreplaceable region", "a haven of biodiversity, without any equal on this continent", "the last primitive forest in Europe", "one of the very rare vestiges of the primary forest in Europe, a kind of living relic of the Hercynian forest", "last vestige of the great temperate lowland forest that covered the whole of Europe", etc.

But what do these words express? What are the precise ecological characteristics that make this forest a baseline model, a showcase site (Sabatini *et al.* 2018; Jaroszewicz *et al.* 2019) (see Chapter 10 of this book). For most authors, it is not the pristine nature of any impact that is emphasized (if not still sometimes as a slogan or media hook). The interest in this forest stems first from the existence and extent of spontaneous biophysical processes that take place within the reserved part of the forest massif. Indeed, such a degree of non-intervention and free evolution on vast surfaces is very rare in the lowland forests of Europe. Moreover, the exceptional antiquity of the situation is underlined: a continuous forest cover for nearly 12,000 years, the absence of visible traces of intervention, etc. Second, this forest is home to emblematic species such as the European bison, the elk, and also carnivores such as the gray wolf or the boreal lynx (see Figure 2.4A). This makes it possible to observe: (i) a network of interactions, particularly trophic interactions, that is more complete and complex than in managed forests, (ii) original structures and compositions of fauna and flora, and (iii) a very high diversity of microhabitats (see Figure 2.4B). Among these original features, the presence of old trees and large quantities of dead wood (and associated species: fungi, saproxylic invertebrates, etc.) is also regularly highlighted. All of these elements participate in making Białowieża an open-air laboratory for biology, ecology and forestry, especially on the issues of naturalness and spontaneity of ecological processes.

Figure 2.4. *The Białowieża Forest is home to iconic species such as the gray wolf (left), a high diversity of micro-habitats and significant amounts of dead wood (right) (photo: L. Godet, May 2017). For a color version of this figure, see www.iste.co.uk/godet/baselines.zip*

Beyond these remarkable naturalist elements, the case of Białowieża is also interesting because this forest is part of an imaginary conceptual model and a field of complex socio-economic relations (Blicharska *et al.* 2020; Barraud 2021). First, it is a fine example of an archetypal baseline, of a showcase, whose exceptional character is regularly mobilized in NGO leaflets (such as the WWF) to call for its protection, in management guides to compare the state of a site and in scientific papers to highlight its heuristic interest (see Figure 2.5). Second, the Białowieża Forest is regularly the subject of controversies that seem to mix scientific arguments and socioeconomic issues. One of the questions that often comes up concerns the natural dimension of this forest and all the implicit issues that this entails: is it a primary forest, virgin, undisturbed, etc.? What is really the nature and degree of its anthropization? However, for some authors, these questions are in fact secondary. Indeed, if its pristine character is a strong issue of communication and identity, it is mostly admitted that this forest is co-constructed by biophysical and social processes. The latter, however, weighs much less than in most comparable woodlands at the European level. Moreover, the naturalness of Białowieża is used by different actors to argue for either stronger protection (*e.g.*, discourses on a wilderness to be preserved for environmentalist NGOs) or for the exploitation of certain trees or plots (discourses of economic or political actors relativizing the virginity of the forest) (Blicharska *et al.* 2020).

Figure 2.5. *The Białowieża Forest as an object of environmental mobilization (source: Marco Lambertini's Twitter account, WWF executive director, April 2017)*

2.4.2. *Characterization of the ecological status of rivers in the United States*

After presenting an example of an approach based on a site that is certainly exceptional, but unique, reducing its universal scope, we now propose an example in complete contrast, in which scientists have sought to establish spatial baselines that integrate all natural variability on a continental scale. Herlihy *et al.* (2008) constructed biological baselines to evaluate the ecological conditions of rivers based on macroinvertebrate assemblages for the entire United States. However, this indicator is particularly sensitive to natural variations in environmental conditions. The constraint in finding baselines sites is thus twofold: (i) these sites must first characterize the very significant natural heterogeneity at this (near-continental) scale and (ii) the degree of landscape and stream alteration within different eco-regions of the United States varies considerably, so that the availability and overall quality of baseline sites also varies among eco-regions. The challenges in this study are first, to discover and select enough sites to robustly characterize the range of settings, and second, to determine a good compromise between simplifying the natural complexity (by regionalization) and defining geographic areas large enough to include enough statistical individuals, while maintaining ecological coherence. Finally, due to the high variability of pressure at this scale, the quality of baseline sites may vary between eco-regions. Thus, some baseline sites in regions with high pressure could be considered degraded in less developed regions. The ecological assessment by comparison to baseline sites is then distorted at the national scale, *i.e.*, sub-continental in the case of the United States of America.

This project shows both the complexity of the issues raised at a scale of analysis covering vast areas, and the methodological difficulties encountered in collecting and homogenizing multiple and substantial data sets. The project presented by Herlihy *et al.* (2008) thus required four years of data collection and then proceeded with the selection and characterization of baseline sites schematically, in three stages: (i) statistical and manual selection of sites representative of each eco-region surveyed, (ii) sampling of benthic macroinvertebrates and collection of water quality and habitat data on these sites, and (iii) application of filters/screening on these data in order to discard and rank the sites according to their level of biophysical integrity. The thresholds for the eliminatory values of the criteria in each of the filters vary according to the eco-region considered. In the end, 1,392 sites were selected (out of 1,655 previously identified) to characterize baseline conditions in nine eco-regions.

2.5. Conclusion

Without opposing the definition of the historical baseline, the spatial approach brings a complementary view and is relevant to support management and restoration actions. This contribution is particularly evident in the constitution of regionalized baseline models, which highlight significant differences in functioning, structures and relationships from one region to another. The space and the situation of the objects have a meaning, and it is possible to integrate them into the baseline notion, modeled or not.

As we have seen, the search for spatial baselines is not without its difficulties (White and Walker 1997): (i) non-existence (or too limited a number) in certain contexts of sites defined on the basis of criteria that are too theoretical or too demanding, (ii) difficulty in defining the right ratio between a desire to embrace diversity (spatial and systemic) and the feasibility of sampling, and (iii) positioning of the threshold relative to anthropic pressure, which refers both to the applied issue of the ambition of the environmental management measures to be carried out, and to fundamental questions about the positioning of humans in their environment, etc. In a pragmatic way, the methodologies implemented today propose many compromises to overcome these limits. One of the crucial points of vigilance emphasized by spatial approaches concerns the importance, when defining the criteria for selecting baseline sites, of not forgetting, according to the principles of hierarchy and panarchy, that a site is necessarily part of higher scale biological and geomorphological systems (Allen and Star 1982; Holling *et al.* 2002; Roche *et al.* 2005). It is then fundamental that the assessment of pressure levels in particular incorporates this nesting of scales (White and Walker 1997; Moss 2015). For instance, it may be inappropriate to select reaches that are certainly without visible direct pressure (dams, weirs, bank protections, etc.), but subject to significant pressures emanating from the watersheds (agricultural, industrial pollution, etc.), or seemingly untouched forest patches, but subject to the influence of the surrounding land use, as a baseline.

Similarly, far from being excluded, the temporal dimension should be integrated into this spatial approach. Indeed, two sites that are comparable at a given moment may follow different trajectories in reality. In this case, the baseline is only valid for a limited period of time, either because the contexts quickly become incomparable, or because the good quality of certain criteria only corresponds to a transitory stage of the site. This is a particularly complex challenge to take up, insofar as we consider that environments are rarely stable and that most of them evolve, in phases (or "stases", to use the panarchic vocabulary) of constant adaptation to global changes.

Finally, this reflection on the spatial baseline also highlights the need to clarify the intended use of each baseline established. It may be a matter of building a baseline model against which to compare a site being studied in order to assess its degradation level, or to define the actions to be implemented to reach this baseline. But, in a less directly applied perspective, an exceptional baseline site can also contribute to feeding imaginary, conceptual models (ecological succession and forest mosaic in Białowieża; longitudinal distribution of biomorphological patterns on the Tagliamento River), and thus feed archetypal representations capable of modifying the socio-cultural frameworks of management measures. If the two objectives are confused, it is possible to fall into a very normative vision that risks homogenizing practices and environments. This is the case, for example, in certain stream restoration practices, where the meandering stream model is most often preferred from a functional and aesthetic point of view, even though it means that it is not consistent with the biophysical conditions of the site (Kondolf 2006) and creating standardized systems (Lespez and Germaine 2016). To address some of these issues, it now seems important to pay particular attention to the study of less well-known contexts by turning to sites that are a priori less attractive from a scientific point of view, but that often cover larger contexts and help to broaden the range of possible baselines.

3

Mapping What is Left of Nature

3.1. Preamble

Here is a child's question: "Are there still places on Earth without humans and where everything is as it was before the expansion of our species?" Do you want to make the child dream and engage in discussions that will take you far into the night? Or are you thinking about your novel, which is waiting for you on the armchair by the fire and invites you to respond a little hastily?

Let us first suggest that the suspense of the novel is unbearable. "No, everything is affected directly or indirectly by humans. If only because of climate change, largely of anthropogenic origin, which modifies all of the ecosystems." You will also evoke Philippe Descola's research and tell the child that even the dense Amazonian forest, which is falsely called "pristine forest", is or has been inhabited and modified by indigenous peoples, in particular by slash and burn agriculture. You will also explain that naturalists are often surprised to find a plant or an insect thousands of kilometers away from its original distribution area, but that it "does not belong there" because it was brought there by the sole of a shoe, a car or a boat and then drifted a little further. You can then conclude that the adventurers, no matter how far they go, will hear a motor noise, observe a halo of artificial light pollution, or see an airplane or the flicker of a satellite in the sky.

But the answer does not seem to satisfy the child's insatiable curiosity. It is rather this lack of perspective in the world that surrounds them that seems to distress them. Therefore, go for a little story that, all things considered, will perhaps help them to dream and therefore fall asleep:

Chapter written by Laurent GODET and Adrien GUETTÉ.

– Do you know the Gangkhar Puensum which culminates at 7,570 meters altitude in Bhutan? You would only know about it from books, because nobody has ever climbed it! Believing perhaps to find extra-terrestrial life forms, humans walked on the Moon, more than 384,000 km from our Planet, before discovering in the 1970s that there were unknown life forms on Earth only a few kilometers deep, within the hydrothermal springs of our oceans, in places deprived of light and oxygen.

But the child has been to school. They know that there are satellites permanently circling our planet, that airplanes take pictures of the Earth from the sky, that robots are sent to the bottom of the oceans, that acoustic sonars detect the shapes of underwater reliefs, that samples are taken from the highest altitudes or the deepest depths. How on earth can there still be unexplored areas?

You will reply that it is not because we have increasingly sophisticated means of observation and sampling that these spaces are inhabited and frequented by humans on a daily basis, nor even that we know absolutely everything about them. A classic satellite image of a dense forest in the Congo Basin informs us mostly about its canopy, not about what is on and in the ground. And you will not resist the temptation to quote Monod (1992), who reminds us how fragmentary the exploration of our oceans is and gives us only a partial image:

> Imagine what one could know of the fauna of France for having explored it only from a balloon, through a permanent and thick layer of clouds, by means of a grappling hook and a salad basket swung blindly at the end of a string. What would we have caught, and still with luck, after fifty years, or even a century? Not much, I'm afraid: a rooster, a few tree branches, one or two pine cones, a Breton headdress, an Alsatian baby, a bra, some oyster shells [...], a very modest anthology. And to which, for lack of a shot glass, a bank bill, a machine gun and a ballot paper, essential elements of our civilization will be missing.

All in all, you are in for a slightly longer presentation than expected. And you will have to sort it out so as not to mix everything up: knowledge, exploration, presence or influence of humans on Earth. The question does not seem as simple as at first sight! Let us focus the presentation on the presence of humans on Earth and their influence. We will have to show that the Earth, even though it seems almost entirely populated and modified by humans, is not uniformly so. There are places that are almost devoid of humans and their sound and visual manifestations... and the methods to bring this to light are quite diverse!

3.2. Introduction

In the context of the Anthropocene, we can consider that the entire planet Earth is under the influence of humans. When a molecule of CO_2 is emitted by our industrial activities from one point of the planet, the whole atmosphere is affected. However, it seems equally obvious that the influence of humans is not uniform. If only for wooded areas, we cannot clearly put the dense forests of the slopes of Borneo, the Landes forest in France, the Siberian taiga, the Amazonian pioneer front and the redwood forests of California into the same category. There are therefore gradients of anthropic influence, and the identification and mapping of more or less natural areas (or their counterparts, *i.e.*, more or less anthropized areas) is the subject of a literature that is both old and fairly abundant.

The stakes of such mapping are numerous. Since human pressure is increasing, the identification of the last natural areas, or at least the spaces that are the least artificial, seems to be decisive for several reasons. First, and this is the main purpose of this book, because these spaces can be considered as non-anthropized baseline spaces (see Chapter 2), spatial witnesses of a temporal phenomenon of anthropization. In this sense, they are space-for-time proxies for which we can make the hypothesis that they attest to what other spaces may have become if they had not been anthropized. The second major challenge is to identify spaces that should be protected as a priority, because they can be considered as remnants of nature that have escaped the steamroller of anthropization. Finally, the last challenges of such mapping are heuristic. What sources and what data should be used to quantify the hold of humans on the Earth? What is meant by "anthropization": is it a matter of mapping spaces that have not been modified by them, that function without them or that are simply far from them? Can we identify clear gradients or breaks in the human[1] footprint? Finally, what temporal and spatial scales should we work at?

We will deal here with maps of areas of nature in its broadest sense (including the notions of wilderness, wildness or naturalness – see Guetté *et al.* (2018) for a definition of the different facets of areas of nature) and its counterpart: maps of anthropization also in the broadest sense (including the diverse and polysemous notions of ecological footprint, human footprint, human pressure or human disturbance). We have chosen to retain the original terms used by the authors as much as possible. Table 3.1 briefly reminds us of what is most commonly referred to by the different terms.

1 See, for example, Lespez and Dufour (2021) regarding the ontological dimension of this gradient.

Terms	Simplified definitions
Areas of nature	
Wilderness	From the 20th century onwards, wilderness areas have come to be defined as opposed to artificial areas. They are original areas, untouched by any transformation by humans. They correspond to a state of unmodified nature.
Wildness	The term wildness refers to the idea of a "quality" of nature that is assessed along a gradient. The spaces with the highest degree of wildness are wildland or wild areas.
Wild areas	Spaces with a high degree of wildness.
Naturalness	The notion of naturalness is also in most cases a "quality", close to and sometimes intertwined with the notion of wildness. Authors often favor the notion of wildness to qualify the degree of freedom (in relation to human activities) of landscapes, whereas naturalness is more often used to characterize an ecosystem in relation to a historical baseline.
Anthropization	
Ecological footprint	Theoretical notion that proposes to measure the surface areas necessary to satisfy the needs of their inhabitants according to their way of life.
Human footprint	Continuum of human influence over the entire land surface. The approach is spatial and aims to identify areas that are more or less marked by humans. Cumulative anthropogenic pressures are one method of measuring the human footprint.
Human disturbance	The term is also used in the sense of human footprint, *i.e.*, the degree of human influence. Initially measured in terms of vegetation, it was later supplemented by other criteria of human presence.

Table 3.1. *Proposed definitions of English-language terms related to areas of nature and anthropization used in this chapter*

In this chapter, we propose a non-exhaustive review of the main approaches to mapping anthropization, their main results in terms of identifying the most natural areas, and the issues and applications of such maps in terms of identifying useful baseline sites for conservation purposes. The divisions proposed in the different mapping approaches do not follow a strictly chronological order and are not perfectly watertight categories. Rather, they are the result of both thematic and methodological approaches that are quite complementary and have largely fed off of each other. We will not present here the history and concepts of wilderness, wildness, naturalness and anthropization as such, but will focus on the major approaches that have proposed maps and methods for locating and identifying areas of nature.

3.3. Zoning of spaces of perceived wilderness: the wilderness of some is not that of others

As mentioned in the introduction, the notion of naturalness is very polysemous, including in its sole acceptance of wilderness. For Nash (1993), one person's wilderness is another person's picnic area, and it is above all what a human thinks it is. It is undoubtedly to show a different appropriation of this notion according to individuals (or, if we see it in a slightly more cynical way, to get around this problem) that the very first wilderness maps explored the wilderness perceived by different groups of actors.

Robert C. Lucas, an American geographer and economist working for the Forest Service in St. Paul, Minnesota, published what can be considered the very first maps of wilderness in the same year that the federal Wilderness Act was passed (Lucas (1964) – see Figure 3.1). He used one of the texts that preceded the Wilderness Act[2] to define wilderness as follows: "a wilderness, in contrast with those areas where humans and their own works dominate the landscape, is hereby recognized as an area where the earth and its community of life are untrammeled by human, where humans themselves are visitors who do not remain". He decided to work outside of the areas identified as such by the American administration (notably the National Forest System) and focused on what he defined as a space of "semi-wilderness", where a few human activities are authorized. He worked on the Boundary Waters Canoe area in northeastern Minnesota. Through user surveys, he identified four different groups of tourists who frequent this area, with canoeists and motorboat users at the two extremes of his typology. The criteria used by canoeists to define wilderness are much more restrictive than those using motorboats. The use of motors[3] even destroys the wilderness of the former. The result is a roughly concentric interlocking of canoeists' wilderness areas in small cores, surrounded by larger wilderness areas of motorboat users. Lucas (1964) proposed zoning space management by concentrating motorboat uses in peripheral areas to maximize the extent (and enjoyment) of space used by canoeists. This pioneering study inspired other studies in New Zealand and elsewhere in the United States in terms of mapping perceived[4] wilderness.

2 This is a 1961 Wilderness Bill.

3 Lucas (1964) does use the term *destroyed*.

4 As well as many other tourism planning studies, which we do not present here.

Figure 3
THE AREAS CONSIDERED "WILDERNESS" BY AT LEAST 50 PER CENT OF THE VISITORS IN EACH OF THE FOUR MAJOR USER TYPES

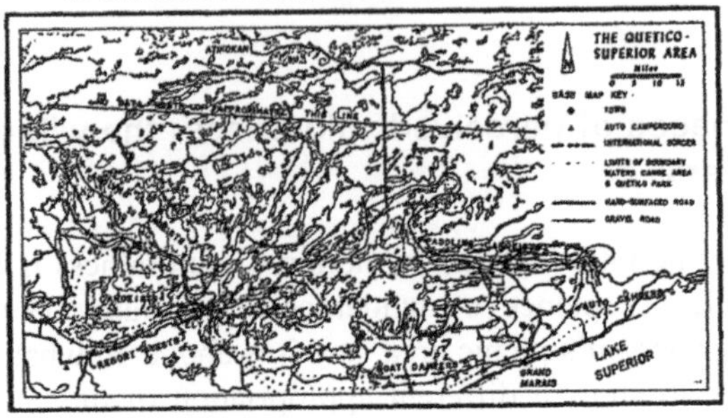

The area in the interior—that is, away from the roads and generally to the north of the line for each user type—was rated as "wilderness" by 50 to 100 per cent of the visitors of that type reaching the area. The dotted portions of the lines indicate data were lacking, and subjective estimates have been made, based on 1960 data.

Figure 3.1. *Although very difficult to read, Lucas (1964) offers one of the earliest maps of perceived wilderness in the United States from the 1960s (source: Lucas (1964))*

The New Zealand geographers Kliskey and Kearsley (1993, 1994) also proposed maps of wilderness perception using a survey method based on the example of a region in New Zealand. They called their method "MWPM" (for Multiple Wilderness Perception Mapping) and asked respondents to rank 16 elements indicating the presence or influence of humans[5] according to five levels (very desirable = 1 point, desirable = 2 points, neutral = 3 points, undesirable = 4 points, very undesirable = 5 points). They obtained a requirement level of each interviewee by summing the points of each element (each interviewee was scored from 16 = minimal level of requirement corresponding to 1 point for each element; to 80 = maximal level of requirement corresponding to 5 points for each element). The authors arrived at four groups identified according to their level of requirement to define wilderness: the strong purists (66–80 points), the moderate purists (56–65), the neutrals (46–55) and the non-purists (16–45). Once the 16 items were mapped, the authors then drew maps of what wilderness was according to each of the four groups. As in Lucas' (1964) study, a map of wilderness emerged with interlocked areas, with the wilderness areas of the purists occupying small cores, while the wilderness areas of the others extended into peripheral haloes. Kliskey (1994) repeated much the same study with some refinement of methodology, and five years later used this study for tourism planning purposes (Kliskey 1998).

5 Campsites, presence of exotic species, road access, commercial recreation, maintained trails, stream crossings, hunting, logging, possibility of motor vehicle crossing, maintained huts and shelters, presence of electrical infrastructure, presence of mines, not seeing anyone else, distance from populated areas, area large enough that it cannot be crossed in less than two days.

Flanagan and Anderson (2008), Australian geographers, largely took up the methods presented above (as well as similar elements to define wilderness) and applied them to the San Juan National Forest in Colorado. They also observed the same interlocking of wilderness spaces between strong, moderate and neutral purists.

3.4. Locating the last wild spots: where is there any baseline nature left?

With the availability of large-scale cartographic datasets, but especially satellite imagery and the democratization of computer processing, the 1970s and 1980s saw the emergence of a completely different kind of cartography: one that identified the last remaining areas of nature on the basis of several indicators that were intended to be more objective.

We can first identify an Australian school of thought. This one being in line with the American currents presenting the wilderness as spaces of quietude outside the civilized world and which have an aesthetic and spiritual dimension. It was even the desire to protect vast spaces for "rugged outdoor activities" and in spaces "devoid of roads or even paths" (Mosley 1992) that led to the first wilderness areas being defined in southeastern Australia. While the first proposals to designate wilderness areas in Australia date back to the 1930s (see Mosley (1992) for a detailed history), the passing of the National Parks Act by the Australian Parliament in 1967 marked a turning point in the history of Australian wilderness. With this legislation, all or part of a park can be classified as a wilderness area. Proposals for mapping wilderness areas, supported by a detailed methodology, began to appear in the 1970s. Helman *et al.* (1976) proposed wilderness mapping of parts of New South Wales and Victoria. Using topographic maps and satellite imagery, they identified and mapped wilderness areas as areas free of human disturbance (notably free of motor vehicle roads and power lines), with core areas of 25,000 ha, free of any indentation, at least 10 km wide and themselves surrounded by buffer zones of at least 25,000 ha. In the years that followed, other authors roughly repeated this same approach elsewhere in Australia (Stanton and Morgan 1977; Feller *et al.* 1979; Russell *et al.* 1979). Kirkpatrick (1979) also developed an approach in Tasmania by taking into account criteria of remoteness (from the first point of access by a motor vehicle) and "primitiveness" (understood as the distance to any anthropogenic disturbance and the possibility of seeing it on the ground). Wilderness cores are defined by Kirkpatrick (1979) as access points with a minimum of a four-hour walk (half a day) and a minimum distance to any form of disturbance that may interfere with a user's wilderness experience. These studies formed the basis for several other studies that later led to the definition of the wilderness continuum concept used in the National Wilderness Inventory. Lesslie, along with various colleagues, became a major figure in wilderness mapping in Australia. Lesslie and Taylor (1983,

1985), then defined a methodology for assessing the quality of wilderness based on two major attributes: remoteness and what could be called "biophysical naturalness" (see Guetté *et al.* (2018)). After producing maps manually, Lesslie and Mackey (1988) and Lesslie *et al.* (1988) proposed computer-based methodologies still based on remoteness and biophysical naturalness indicators (remoteness from any human settlement, remoteness from road access, aesthetic naturalness determined by the absence of anthropogenic structures in the landscape, biophysical naturalness understood as the degree to which the natural environment is free from disturbance or influence of modern human technology).

The leap in scale was made a few years later by McCloskey and Spalding (1989), two American members of the Sierra Club, who proposed a map of the wilderness on a global scale (see Figure 3.2). Their approach was clear: the last wild lands are a world heritage of humanity and the study they propose is intended to be a report to prevent it from being "nibbled away with impunity". They accepted a definition of wilderness not as untouched areas (which they claim no longer exist), but as areas without permanent human settlement, without roads, without crops or regular grazing. While being aware of the imperfect character of their study, they used two sets of maps from the US Defense Mapping Agency: the Jet Navigation Charts (2,000,000th scale) and the Operational Navigation Charts (1,000,000th scale). These maps are used for military and aeronautical navigation purposes and provide land cover information on a global scale. In order to identify wilderness areas, they identify all areas with the following features: roads, buildings, airports, railroads, pipelines, power lines, canals, pavements, aqueducts, major mines, dams, reservoirs and oil wells. Only areas without these elements and of a minimum of 400,000 ha were retained. Their map shows that about one-third of the world's land mass (excluding Antarctica) is wilderness and two-thirds is dominated by humans. It is these areas that should be conserved as a priority.

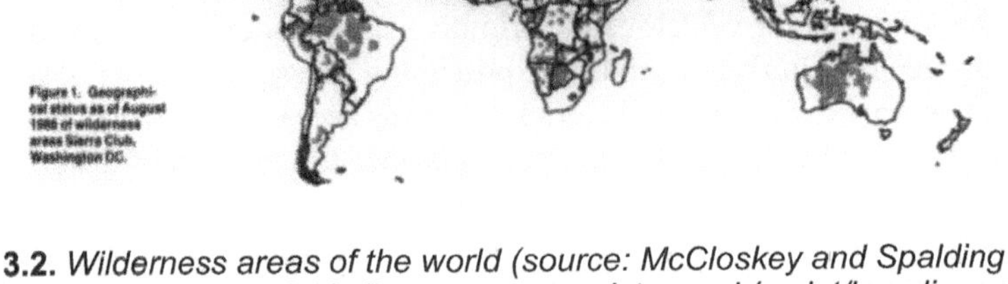

Figure 3.2. *Wilderness areas of the world (source: McCloskey and Spalding (1989)). For a color version of this figure, see www.iste.co.uk/godet/baselines.zip*

While it is usual to attribute the paternity of the concepts and the implementation of the conservation of large natural areas to American or Australian authors, it is necessary to note that Russia also conceptualized the idea of identifying and protecting large nature sanctuaries very early on. This is reflected in the work of the geographer and biologist Vasily Dokuchaev and the theory of *zapovednost*, which is based on the need to prohibit all human activity (except for incursions by scientists and forest rangers) in certain sites. *Zapovedniks* are presented as parcels of intact natural ecosystems that are benchmarks, witnesses, to be compared with anthropized ecosystems (for a history of *zapovedniks*, see especially Chibilev (2017)). They must be large enough to be self-regulating. Despite this long history, the first Russian mappings based on a standardized method were only published in international journals in the 2000s. Yaroshenko *et al.* (2001) propose maps of so-called "intact forests" of European Russia. Working on this space is interesting because it is a space which appears, in its northern part, as a wilderness zone according to McCloskey and Spalding (1989) (and also according to other studies which appeared later) and because it is a full ecosystem. Thus, there can be cores of nature within the same ecosystem, whereas the approaches used up until now were mainly concerned with administratively delimited spaces or the whole planet. The whole forest ecosystem is taken into account (wooded areas, and also their wetlands, rivers, lakes and open areas). The authors' goal is clear: to identify baseline forests that will allow a better understanding of the functioning of forests disturbed by humans (hence the need to take into account large blocks of forest ecosystems) and to identify what should be conserved. Yaroshenko *et al.* (2001) consider intact forests to be areas of at least 500 km^2 and at least 10 km on a side, with no evidence of permanent human occupation or communication, industrial logging in the last 60 years, mining or clearing or other significant human impacts. The methodology builds on elements seen previously and proceeds in three steps: (1) from maps: elimination of large disturbed areas (cities, roads, etc.); (2) from 150 m resolution satellite data: identification of tundra areas, logging, agricultural areas, etc.; (3) from Landsat data and similar 30 m resolution satellite images: identification of other disturbances. A ground truth survey was conducted on 67 areas. The authors showed that 14% of the Russian European forest corresponds to large blocks of intact forest, which are mainly concentrated in the Far North and in peatlands. A few years later, the same authors, accompanied by 12 new co-authors, renewed this approach on the forests of the world (Potapov *et al.* 2008). Slightly more than 23% of the forest areas, mainly in dense tropical and subtropical areas (45.3%) and boreal forests (43.8%) are considered as intact.

3.5. Nature areas broken down into facets and gradients: are there tipping points in space?

As previously mentioned, the United States, legislation, specifically the Wilderness Act of 1964, has defined a type of nature area, understood as wilderness. Wilderness areas included in the Wilderness System, which today covers nearly 430,000 km² and is administered by four federal[6] agencies, have thus been identified and mapped. In Australia, the chronology and approach are very similar and took shape with the passing of the National Parks Act in 1967. In Russia, the ancient idea of *zapovednost* led to the designation of nature sanctuaries: the 101 current *zapovedniks* cover 330,000 km² and are almost all managed by the Russian Ministry of Natural Resources. In Europe, on the contrary, it was not until 2009 that the European Parliament passed the Resolution on Wilderness (Parlement Européen 2009), which aims to: (1) better define wilderness, including ecosystem services and conservation values; (2) set up a mapping program to identify the remaining wilderness areas, the current distribution of biodiversity and places where human activity is minimal and (3) promote better protection of wilderness areas from threats. This resolution has led to a great deal of work. Since it is more recent than the Russian, American, Australian and New Zealand work, the mapping of nature areas in Europe started right away with the use of fairly comprehensive geographic databases and benefited from all the modern tools of geomatics. As we will see, the approach is quite different in that it clearly proposes to map in a rather fine way not only **gradients** of nature, but also different **facets** of this nature. The very idea of gradients of nature, identified quite early by several authors, can now be quantified and mapped in Europe. Nash (1993) considers that wilderness is indeed a gradient "from paved to the primeval". These gradients have been the subject of much debate as to the tipping point at which one enters wilderness (Lesslie and Taylor 1985; Nash 1993; Carver 1996; Dawson and Hendee 2009). However, with such approaches, any space can be mapped to identify nature areas, including both lightly and intensely anthropized spaces.

Steve Carver, a geographer from the University of Leeds in England, accompanied by various collaborators, can be considered as one of the pioneers of wilderness mapping in Europe (see Carver (1991); Carver and Fritz (1999)). One of the original features of his work was to produce gradient maps of different facets of wilderness, using the processing of different databases by geomatic tools, mainly multi-criteria analyses (Carver was one of the first to implement such analyses in Geographic Information Systems (GIS) – see Carver (1991)). Carver *et al.* (2002) proposed wilderness maps at the scale of Great Britain based on six criteria (distance to local population cores, to national population nuclei and to access to the road

---

6 National Park Service, U.S. Forest Service, U.S. Fish and Wildlife Service, and Bureau of Land Management.

network, distance to visible human infrastructures, degree of naturalness of the land cover, altitude). A wilderness map is produced from these criteria according to a multi-criteria analysis using a linear weighting of each criterion. On an online web interface, each user can assign the desired weight to each of the six criteria to generate a wilderness map (see Figure 3.3).

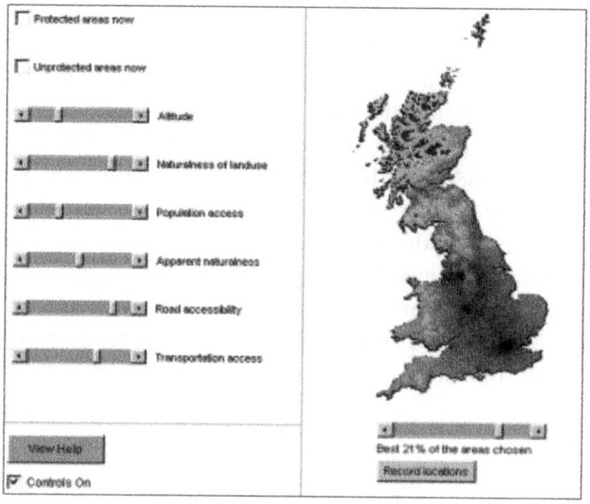

Figure 3.3. *Web interface to wilderness mapping in Britain (source: Carver* et al. *(2002))*

With Comber *et al.* (2010), Carver also applied multi-criteria analyses to map wildness gradients in Scottish national parks (Comber *et al.* 2010; Carver *et al.* 2012). The criteria considered were those usually found in the literature: land cover classified in degrees of naturalness, distance to human infrastructure, roads and inhabited areas and topography data. For the authors, the identification of these important spaces is crucial for identifying recreational spaces, and also as iconic landscapes of Scotland that are linked to the national identity. When Carver *et al.* (2013) developed wilderness maps for the Death Valley site in the United States, they focused more on individualizing four different **facets** of wilderness corresponding to the definition of wilderness in the Wilderness Act (natural quality, untrammeled quality, undeveloped quality, solitude or primitive and unconfined quality), and the synthetic wilderness map is simply the sum of the values of these four facets.

Such studies were then conducted in different European countries: in Italy at the Dolomites UNESCO site (Orsi *et al.* 2013), and even in Denmark where

Müller *et al.* (2015) took up the challenge of mapping the "wilder parts" in a country they knew was already highly anthropized[7]. These maps are also being used by authors to assess rewilding opportunities on a European scale (see Ceaușu *et al.* (2015)).

Maps of nature areas are now available for Western[8] Europe. Using and adapting the methodology by Carver *et al.* (2012), the team at the Wildland Research Institute (University of Leeds) has developed a spatially explicit index of wilderness (Wilderness Quality Index) at the scale of Western Europe (see Figure 3.4). Measured on the basis of four attributes: distance to roads/rail, population density, land use and landscape ruggedness, the mapped result identified the wildest areas (the "top 10%"). Not surprisingly, these results had a clear tendency to be marked by altitudinal (high altitude spaces) and latitudinal (high latitude spaces) factors. However, this time these gradients were confronted with other realities, notably political and biological. The authors sought to test the spatial congruence of high wilderness areas with European protected areas and with the distribution areas of protected mammal species. These studies could help to determine thresholds or tipping points of anthropization intensity according to the habitats or species of interest.

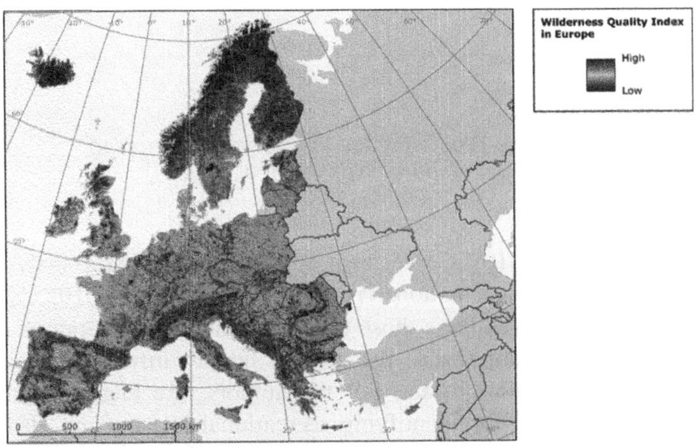

Figure 3.4. *Wilderness quality index map (source: European Environment Agency). For a color version of this figure, see www.iste.co.uk/godet/baselines.zip*

7 Outside of Europe, Hou *et al.* (2019) also propose naturality gradient maps for the Beijing metropolitan area.

8. See: https://www.eea.europa.eu/data-and-maps/figures/wilderness-quality-index/wilderness-quality-index-in-europe.

3.6. Anthropization of nature: summarizing the influence of humans in a single index

The maps that synthesize the influence of humans are the counterpart of the maps of nature areas. The approach consists of proposing synthetic indices of anthropic pressure at different scales. The terms used, which cover more or less the same thing, are mainly disturbance, footprint or human pressure.

The first studies mapping the influence of humans on Earth use the term **human disturbance**. It is worth mentioning the seminal work by Bennett (1975). For each of the major regions of the globe (sub-Saharan Africa, Asia, Europe and the Mediterranean basin, Australia, North America, South America, island environments and marine ecosystems), Bennett proposes maps of human influence according to three levels of disturbance of vegetation considered as wild: major disturbance (areas where only a few remnants of wild vegetation remain), moderate to strong disturbance (areas with large scattered areas of wild vegetation) and weak disturbance (areas with large areas of wild vegetation) (see Figure 3.5).

Figure 3.5. *Map of human disturbance in Europe (source: Bennett (1975)). For a color version of this figure, see www.iste.co.uk/godet/baselines.zip*

Mapping of human disturbance on a global scale continued with the work of Hannah *et al.* (1994)[9], as shown in Figure 3.6. These authors broke down disturbance into three levels: disturbed, partially disturbed and human-dominated. The criteria used are as follows: vegetation (primary or secondary); human population density; type of agriculture (shifting, extensive, permanent); cities; livestock density and various sources of human disturbance (only forestry and desertification are explicitly mentioned). The second index they proposed is that of the remaining natural habitat. According to them, 52% of the planet remains undisturbed by humans.

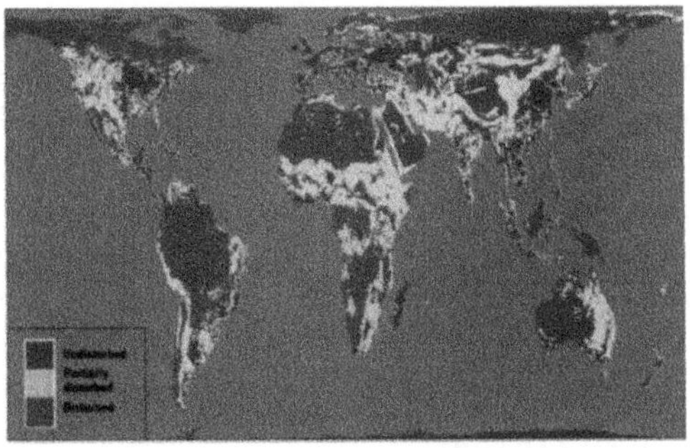

Figure 3.6. *Global mapping of human disturbance (source: Hannah* et al. *(1994)). For a color version of this figure, see www.iste.co.uk/godet/baselines.zip*

The **ecological footprint** is a measure of the surface area required to meet the needs of natural resources and absorb human pollution, on an individual or population scale. This concept, largely developed by Wackernagel and Rees (1996), is applied to cities to determine the surface area required to satisfy the needs of their inhabitants. It is thus obvious that a city, whatever it may be, placed in a form of lockdown and thus cut-off from the outside world, would perish quite quickly. Wackernagel and Rees (1996) point out the imbalances in ecological footprints among humans: a North American has a footprint of 4 to 5 ha, three times the sustainable human footprint (in other words, it would take at least three planets for humanity to live sustainably if all humans lived like a North American). The Dutch, on the contrary, have an ecological footprint equivalent to 15 times the area of their country (Wackernagel and Rees 1996). This is therefore a land-use planning and economic[10] approach. This approach does not really propose maps, except in a very schematic way, such as the representation of the size of a country and the size of its

9 Their method is complemented by Hannah *et al.* (1995).

10 Mathis Wackernagel is a Swiss engineer and planner, director of the Global Footprint network; William Rees is a Canadian professor of urban planning and ecological economics.

ecological footprint at the same spatial scale. The ecological footprint is thus a virtual space, most often represented around an area inhabited by humans.

The notion of **human footprint** appears in an article by Janzen (1998) in the form of an essay reminding us that everything on Earth is more or less anthropogenic, and then in a commentary by Glaze (2001), but without any real definition. Sanderson *et al.* (2002) were, in fact, the first to propose both a definition and a cartographic expression for the human footprint. They claim this because of the need to move away from abstractions, such as the appropriation of net primary production or exponential population growth. They raise the need to understand the spatial extent of human influence in the form of maps. The 1990s made this possible thanks to: (1) satellite data providing reliable land cover and land use data on a large spatial scale; (2) population data on a global scale and (3) the development of GIS to combine all these data. Sanderson *et al.* (2002) are well aware of the similarity of the concept to the ecological footprint and define the human footprint as the sum of the ecological footprints of the human population, expressed as a continuum of human influence over the entire Earth's surface, revealing the distributional patterns of human influence on nature. They also caution the reader that the human footprint does not measure the impact of humans per se, but rather the zones of influence where humans have more or less responsibility for the biological compartment. Sanderson *et al.* (2002) use four types of data (grouping nine different datasets): population density; land transformation (land use and land cover, buffer zones around roads and railroads, built-up areas); accessibility (based on the distance a human can travel from a road, river or coast) and energy infrastructure (areas more or less illuminated by artificial lights). Everything is then presented by a grid of 1 km^2, each dataset is given an identical score from 0 (no human influence) to 10 (high human influence). The result is a general map of human influence. They add the identification of the 10% wildest spaces in each biome and biogeographic province to their analysis. From these spaces, the *Last of the wild* is identified as the ten largest "10% wildest cutoff" areas.

A methodology for mapping human impacts has been proposed by UNEP (UNEP (2001) cited in Sanderson *et al.* (2002)), with an index of human disturbance on fauna and flora as a function of distance from anthropogenic infrastructures. Initially focused on the Arctic, the method was then extended to the whole planet (see: www.globio.info). Numerous variations of human footprint mapping are now proposed at regional scales (see Woolmer *et al.* (2008), who rightly emphasize the interest of rejecting these mappings at regional and local scales to finely identify differences), both on spaces considered fairly intact (Antarctica: see Hughes *et al.* (2011); Pertierra *et al.* (2017)) and much more anthropized (as in Japan: see Yang *et al.* (2014)).

The latest major advances concerning the mapping of anthropogenic pressures were developed at sea (see Halpern *et al.* (2008)) and take into account a temporal dimension. Geldmann *et al.* (2014) thus calculated a Temporal Human Pressure Index (THPI) largely inspired by the index of Sanderson *et al.* (2002), but enriched by 17 additional datasets (the THPI is calculated from human population densities, land transformation and electrical infrastructure) from 1990 to 2021, and compared it inside and outside protected areas. Venter *et al.* (2016) also proposed a map of the evolution (1993–2009) of the human footprint (based on the presence of built-up area, arable and grazing land, population density, artificial night lights, roads, railways and waterways).

3.7. Anthromes: ending the divide between the natural and the anthropogenic?

In recent years, the approaches listed above, along with the wilderness maps and human footprint maps, have been blended. Allan *et al.* (2017) used Sanderson *et al.*'s (2002) last of the wild formula and Venter *et al.*'s (2016) human footprint study to map the "pressure free" areas of more than 10,000 km^2 and the 10% of spaces least subject to anthropogenic pressures in each of the 60 biogeographic regions of the globe. Tricker and Landres (2018), meanwhile, analyzed threats to wilderness areas in the United States according to 14 different indicators.

Riggio *et al.* (2020), after listing all the applications that human footprint maps could have (loss of wilderness, species extinction risk, changes in animal behavior and movement, effectiveness of protected areas, etc.), compared human footprint maps in the tradition of Sanderson *et al.*'s (2002) work, maps of human-modified ecosystems, maps of anthropogenic modifications and maps identifying areas of lowest anthropogenic impacts. The conclusion of this comparison was that regardless of the method, the areas identified as being of high wilderness are roughly the same – namely, arid deserts and polar regions – with total areas of low human impact ranging from 20 to 34%.

Although the study by Ellis and Ramankutty (2008) was used in Riggio *et al.* (2020) and placed on the same level as many others, it proposes an approach that breaks away from all of the maps presented in this chapter. These authors combine, on the one hand, biomes with, on the other hand, human population and land use and occupation. For these authors, considering the planet in terms of large biomes (based on large vegetation types and regional climatic variations) with humans and their activities considered only as a disruptive element to this distribution scheme is non-sense. The authors prefer to speak of "anthromes" (or anthropogenic biomes), of which there are 18, and on the contrary of wild biomes, the latter representing less than a quarter of the ice-free land area. If we interpret their words, there would not really be baseline spaces, but rather an interpenetration of humans and nature. The baselines would rather be sought in the way human societies have lived with their environment. Ellis *et al.* (2021) showed that even 12,000 years ago, nearly three quarters of the Earth was inhabited and thus shaped by human societies, including more than 95% of temperate woodlands and 90% of tropical forests.

3.8. Conclusion

Over the past half-century, the mapping of nature areas (and its counterpart, anthropization mapping) has developed largely around different approaches that have evolved according to the "schools" of mapping, regional contexts and conservation issues. These maps are also marked by the effect of the multiplication of data and the development of new analysis and calculation tools.

In the recent history of nature and anthropization mapping, we can first distinguish so-called dichotomous approaches that tried to divide the space into two: areas occupied by humans and, in contrast, natural areas. These approaches were used very early on in the United States, in Australia, and also in Russia. These spaces without humans are then considered as baselines because of their original and untouched character. They constitute national jewels that are admired for their beauty and past splendor, and in which we can escape from the civilized world.

In parallel, other methodological approaches propose to free themselves from strict and arbitrary categories by developing so-called gradient (or continuum) maps of nature and anthropization. These approaches are more nuanced because they consider the effect of humans everywhere on the planet, but attempt to detect orders of magnitude. It is no longer a question of mapping what is outside of civilization, but what is more or less anthropized and more or less natural. The advantage of this relative approach is that it allows the mapping of all types of spaces, from boreal forests to urban or agricultural spaces, and at all spatial scales, within a city or a continent. Each space is thus measured not in relation to a "natural" baseline, but in reference to all other spaces. However, the counterpart of this relativity is that it does not enable the detection of anthropization and naturalization dynamics. Since everything is relative, these maps can be repeated in any space, at any time step, and we will always find gradients with spaces that are less artificial than others, without detecting the dynamics at work. Such approaches can therefore contribute to environmental amnesia (see Chapter 4).

Refusing to see humans as disruptors of a natural baseline system, we must also mention the existence of approaches that claim the interpenetration of humans and non-humans. This time, it is not a question of classifying space as natural, unnatural or more or less natural (which is somewhat normative), but of distinguishing categories of interweaving between humans and non-humans forming new mosaics. The baseline to a natural state is then abandoned. There would be, so to speak, no more nature.

Despite theoretical and methodological developments, the nature areas and anthropization maps produced today remain imperfect. A certain number of subjective elements are the responsibility of the authors: methodological choices (selection of indicators, aggregation methods, weighting, etc.), theoretical choices (whether or not to take into account humans as a disruptive element of the planet), not to mention the subjectivity inherent in the cartographic object (type of representations, projections, semiology, etc.). Moreover, the massive use of land use data from satellite images does not allow us to apprehend certain forms of environmental transformation (intra-forest degradation, homogenization of facies, etc.).

However, in the context of extremely rapid anthropization of the planet, cartographic tools developed in a rigorous scientific approach (transparency of data and replicability of methods first) offer extremely powerful tools to measure the extent of the changes at work and assist in conservation decisions.

One of the most promising avenues of these mapping approaches is their potential to create baselines. Thanks to the temporal hindsight of the first maps, we are beginning to see the possibility of measuring spatio-temporal evolutions. We thus move from documenting a state to documenting a dynamic. This relative approach makes it possible – in part – to counter criticism of the choice of indicators and thresholds used. Today's maps can be seen as contemporary baselines, like photographs at a given moment, from which we can detect anthropization or, on the contrary, impoverishment dynamics in the future. These tangible data produced and shared by the international scientific community constitute elements of observation that make it possible to fight against the environmental amnesia that, in the absence of a documented baseline, pushes us to move our temporal baseline cursor only a few decades back. If the global maps reveal the existence of more or less anthropized spaces across the planet, the successive maps reveal more or less ancient and more or less rapid anthropization dynamics. These maps of nature areas and anthropization can therefore be seen as temporal milestones that can serve as baselines for the future.

4

The Baseline: A Social Construction

4.1. Introduction

In 1995, the marine biologist Daniel Pauly published his article *Anecdotes and the shifting baseline syndrome of fisheries*. According to his theory, the baseline, *i.e.*, the state of an ecosystem considered as "natural", evolves over time, without the actors noticing it, insofar as they consider an erroneous baseline: that of previous decades, corresponding to their childhood or the beginning of their professional career. As each generation redefines what is natural and what is not, the standard becomes slippery, and the degradation of ecosystems becomes invisible to the actors. He states that the quantity of fish considered normal in a marine ecosystem tends to decrease with the decreasing age of the researchers (Pauly 1995).

There are several reasons why this short account, written in the first person, has resonated widely in fisheries management and marine ecology. The theory is easy to grasp and its logic seems compelling (Campbell *et al.* 2009). It also highlights a hitherto imperceptible cause of ecosystem degradation, linked to our own representations. To consider an ecosystem that, only yesterday, would have seemed "degraded" to us as "natural" is to revise what is desirable in terms of conservation downwards (Maris 2018; Soga and Gaston 2018). The risk is then to become increasingly tolerant of ecosystem degradation (Soga and Gaston 2018). Another strength of this theory is its evocative power: it summons images of past abundance, which contrast with a current degraded state; the theory is then mobilizable for conservation programs (Campbell *et al.* 2009; Alagona *et al.* 2012). Finally, within the scientific community, the analysis of the shifting baseline syndrome makes it possible to encourage interdisciplinary work, since it requires the combination of

---

Chapter written by Clémence MOREAU, Cécile BARNAUD and Raphaël MATHEVET.

historical data not only from ecology, but also from local knowledge, art or literature (Campbell *et al.* 2009).

Through this text, Pauly invites us to take a new look at the concept of a baseline, defined as "a state or trajectory of a system used as a comparator" (Maron *et al.* 2015). Far from being a stable and neutral datum, the baseline is influenced by political agendas, economic realities, scientific knowledge, local knowledge and even prejudices (Hilding-Rydevik *et al.* 2018). The baseline can thus be screened by constructivism, according to which we can only access knowledge of reality through our system of representations, individual or collective (Lévy and Lussault 2013, pp. 224–227).

Considering the baseline as a social construct calls for the exploration of at least three lines of thought that we will develop in this chapter. First, the baseline is not stabilized but evolves over time: this was Pauly's intuition, which has been confirmed but also criticized by numerous studies. Second, different elements, not only ecological, but also social, cultural, economic and even political, enter into the construction of the baseline, hence the fact that the baseline may vary according to social groups or individuals, and to contexts. Finally, since the choice of the baseline has strong management implications and there is not necessarily a consensus, it is necessary to debate it. To develop these three lines of thought, we will rely on a review of the literature and support our remarks with examples taken from a study conducted in the Cévennes National Park (see Box 4.1).

Mont Lozère is a granite massif located in the south of the Massif Central. Its characteristic landscapes (pastures dedicated to sheep and cattle breeding, natural meadows, beech/fir forests and coniferous plantations), as well as the cultural and ecological richness of the site, have justified various classifications and recognitions: Cévennes National Park in 1970, biosphere reserve of the Unesco Man and Biosphere program (1984), inscription of the Mediterranean agro-pastoral landscapes on the Unesco World Heritage List (2011) (see Figure 4.1).

As elsewhere in France and Europe, the Lozère mountain region is witnessing the "closing of landscapes", *i.e.*, the increase of forest cover in former rangelands and meadows, abandoned or less intensively grazed. In these mountain regions, the expression "landscape closure" implicitly refers to a negative, undesirable phenomenon (Le Floch *et al.* 2005). One of the structuring postulates of the fight against "landscape closure" is the existence of a baseline landscape, situated in the middle of the 19th century, when the agricultural exploitation of the territory was at its peak and the landscapes were particularly open. The increase in forest cover is then perceived as a deviation from this model (Marty and Lepart 2001; Le Floch *et al.* 2005). At the same time, the open landscapes themselves are evolving, in the context of a (moderate) intensification of livestock systems, and there is disagreement on the types of open landscapes that should be preserved or promoted. Certain agricultural practices are indeed on the increase, such as rock removal (which consists of removing large granite boulders to facilitate the mechanical exploitation of plots) or plowing of meadows (the plowing of permanent meadows or rangelands, which then become temporary meadows, plowed and seeded regularly). Insofar as they may impact the character of the park, these two practices are subject to authorization in the core zone of the National Park.

There may therefore be tension between the baseline landscapes (extensive agro-pastoral landscapes) and the current landscapes (which are evolving either towards forest or towards more intensive agricultural landscapes). This tension manifests itself in conflicts of interest, or even open conflicts, notably between the agents of the Park and the farmers around certain agricultural practices (rock removal and plowing of meadows).

To study this tension, we set up a qualitative research device between 2015 and 2019, as part of a PhD thesis in geography, combining a qualitative approach (semi-structured interviews, participant observation) and action research (participatory workshops around a role-playing game) (Moreau 2019).

Box 4.1. *Presentation of the case study*

Figure 4.1. *Agro-pastoral landscapes of Mont Lozère (source: N. Salliou). For a color version of this figure, see www.iste.co.uk/godet/baselines.zip*

4.2. The baseline evolves over time: the shifting baseline syndrome

The shifting baseline syndrome theory has met with widespread acceptance in the scientific community. A meta-analysis, conducted in 2019, indicates that 152 publications have been made on the topic, 82% of which are on aquatic ecosystems and 28% on terrestrial ecosystems (Guerrero-Gatica *et al.* 2019). On the contrary, the appropriation of this theory is still timid in the French-speaking literature. Changes in baseline systems are essentially addressed by the social sciences, which question the representations underlying the development of land-use planning policies (Luginbühl and Terrasson 2013), but without explicitly referring to the shifting baseline syndrome.

In the scientific literature, three types of papers refer to the shifting baseline syndrome: those that reconstruct a past baseline by measuring species abundance, those that focus on representations and finally those that address the management implications of this syndrome.

The first type of paper aims to assess changes in the abundance of species, mostly marine. All of these papers agree on ecosystem degradation, but they mobilize a variety of resources and methods (Pinnegar and Engelhard 2008). Some authors rely on scientific or naturalist data (Baum and Myers 2004; McLean *et al.* 2016), while others use fisheries catches and trade data (Guénette and Gascuel 2012; Alleway and Connell 2015). Some studies mobilize less formal sources, such as historical photographs of recreational fishing trophies (McClenachan 2009) or archeological finds (Drew *et al.* 2013). Other studies rely on both formal and informal data (Godoy *et al.* 2010; Whipple *et al.* 2011; Fortibuoni *et al.* 2016). Finally, several works seek to incorporate Pauly's cherished "anecdotes" (Ainsworth *et al.* 2008; Parsons *et al.* 2009).

While some studies take a comparative approach, and replicate the same data collection protocols used in studies conducted several decades earlier (Price *et al.* 2014), most of these studies simply paint a picture of past ecosystems, without addressing the issue of perceptions. They evoke a shifting baseline syndrome from the baseline without being able to demonstrate it.

The second type of paper focuses on representations of the baseline and its evolution through time. Several studies compare the representation of a "natural" or "good condition" ecosystem across generations (Bunce *et al.* 2008; Turvey *et al.* 2010; Burbano *et al.* 2014; Giglio *et al.* 2014; Katikiro 2014; Barbosa-Filho *et al.* 2020). Researchers have reproduced the same survey protocol as a study conducted in the 1970s (Leeney and Downing 2016). An ethnobotany study compares the ability of different generations to name local species in their local language (Kai *et al.* 2014).

Finally, in the tradition of Pauly, who took himself as an object, some studies focus on the evolution of the baseline in the scientific life of researchers (Muldrow *et al.* 2020).

These works conclude that there is a change in the baseline between generations, which Papworth (2009) calls generational amnesia. Here again, these articles address the question of representations, without necessarily mobilizing scientific data that would allow us to compare the past state of the ecosystem with the perceived state. All in all, few studies combine a rigorous description of ecological dynamics with an analysis of representations.

Finally, the third type of article focuses on the management consequences of the baseline shift. The shifting baseline syndrome creates an increased tolerance for environmental degradation, resulting in a low level of ambition for conservation or restoration (Soga and Gaston 2018; Guerrero-Gatica *et al.* 2019). Moreover, if the baseline is not perceived in the same way by all the actors, there may be conflicts between those who implement restoration operations (scientists, managers, technicians) and the users (Guerrero-Gatica *et al.* 2019). To prevent this syndrome, several avenues are possible: improving knowledge of past ecosystems, encouraging direct experiences with nature or environmental education (Soga and Gaston 2018). Until then, it is necessary to follow the precautionary principle to avoid underestimating the degradation of the environment (Dudgeon 2010).

Some authors take a critical look at Pauly's theory. Longitudinal studies of ocean history are few, and studies often mobilize unreliable data; they also resort to extrapolation or focus exclusively on certain species (Campbell *et al.* 2009; Alagona *et al.* 2012). The shift in the baseline is also built on assumptions that are not made explicit. The first is the existence of a pristine nature, past and gone, that humans have inexorably degraded. This assumption deserves to be discussed, insofar as it is based on a typically Western separation of humans and nature, according to which marine ecosystems are not populated (Campbell *et al.* 2009). Many protected sites are more cultural than natural landscapes (Hilding-Rydevik *et al.* 2018). Moreover, this vision of a past stability contrasts with a dynamic conception of ecosystems, which prevails in ecology today (Campbell *et al.* 2009). The second unexplained hypothesis is that the past baseline also constitutes a management objective, a desirable future, in a nostalgic approach to conservation ecology. However, Alagona *et al.* (2012) remind us that if the past can be used as a parable, it cannot be used as the only guide for the future.

All of these reflections invite us to pay close attention to the different representations of the baseline, and to the tensions and disagreements that can exist around the shifting baseline syndrome between the actors of a territory. This is what we highlight in the case study of Mont Lozère (see Box 4.2).

We explored the hypothesis of a shifting baseline syndrome on Mont Lozère, based on 46 individual semi-structured interviews, with actors from the agricultural, forestry, tourism, conservation and hunting communities (2016–2017), and participant observation of the Scientific Council of the Cévennes National Park (*Conseil scientifique du Parc National des Cévennes*) (2017), notably a meeting which focused on landscape evolution in relation to agro-pastoral practices (Moreau *et al.* 2019a).

Our results show that among the actors in the field, the dominant representation is that of "landscape closure", experienced as a negative phenomenon, against which it is necessary to fight by "landscape maintenance", *i.e.*, by maintaining "open" landscapes, or even by reopening certain "closed" environments. However, other representations exist, such as that of the foresters, who consider that "we must take advantage of the increase in forest cover", by an ecological or economic development of these landscapes. These actors situate the baseline in a period prior to the agricultural development of the territory, sometimes "the Middle Ages", often in more vague terms. They put forward the "natural" character of the forest, referring as much to the past state of the ecosystem as to a climax. We can therefore speak of a shift in the baseline, insofar as forest landscapes are regaining legitimacy. Indeed, in the past, such normative shifts have already taken place: it is only since the 1970s that the ecological and cultural interest of pastoral landscapes has been put forward (Marty and Lepart 2001).

At the same time, although a large number of actors wish to "maintain open landscapes", they do not all agree on the type of open landscapes to be preserved. For the National Park agents or the tourism actors, the evolution of agricultural landscapes under the effect of rock removal and plowing of meadows constitutes a "degradation" of heritage landscapes. According to them, there is a significant risk of a shifting baseline syndrome: insofar as these changes are temporary and gradual, they will not necessarily be perceptible to the actors who will gradually get used to them. For other actors, mainly livestock farmers, the evolution of landscapes under the effect of these practices is considered inevitable, even desirable, often described as an "improvement" of the land and an increase in its productive potential, which is in line with the practices of their ancestors who shaped the landscapes to practice livestock farming.

We are thus witnessing two joint phenomena: on the one hand, the tipping of the baseline through a questioning by certain actors of the supremacy of open landscapes as a baseline, and on the other hand, a shifting baseline syndrome linked to the evolution of open environments under the effect of the intensification of agricultural practices. These two phenomena are partly linked, as the intensification of practices in open environments can be used by some as an argument in favor of a return to a more natural forest environment. In both cases, mobilizing different baselines is a way for actors to support their practices and their own legitimacy to "create landscapes" (Hatzfled 2009), whether they are breeders, National Park agents, agronomists or foresters.

Box 4.2. *Tilting and shifting of baselines regarding Mont Lozère*

4.3. How is the baseline constructed?

If we accept that the baseline is a social construct, it remains to be understood how this construction takes place. Soga *et al.* (2018) identify several causes of shifting baseline syndrome: incomplete data about the natural environment, impoverished experience and familiarity with the environment. We can therefore identify four factors that are involved in the construction of the baseline: knowledge, memory, experience and cultural norms.

Regarding the first of these factors, several works illustrate how changes in knowledge have sometimes led to a reconsideration of the baseline. Thus, at the end of the 19th century, there was a debate among vegetation specialists about the origin of the open landscapes of the Grands Causses: in the 1880s, the French botanist Charles Flahaut argued that the vegetation was a relic of steppe vegetation, before reversing his position at the end of his career, and asserting that the climactic vegetation of the Causses was forest, with overgrazing being able to explain the current landscapes (Marty and Lepart 2001). These scientific reversals have significant consequences for management. For example, Vera *et al.* (2010) explain that between the 19th and early 20th centuries, the degradation of European forests was attributed to the presence of herbivores (see Chapter 5). On the basis of this evidence, two measures were prescribed to return ecosystems to a "natural" state: spontaneous development of vegetation and exclusion of domestic herbivores. However, more recent studies have shown that domestic animals were integrated into forest ecosystems during the Middle Ages. In addition, scientific knowledge can also be mobilized by actors according to their own strategies, for example by conservation organizations (Barnaud *et al.* 2021) or agricultural unions concerned with maintaining or developing certain practices (Mathevet and Béchet 2020). Finally, alongside scientific knowledge, local knowledge can also be taken into account in the construction of the baseline. In some cases, the erosion of local knowledge can explain the shifting baseline syndrome (Kai *et al.* 2014). Finally, we can say that knowledge is socially and historically situated: even if the baseline is based exclusively on knowledge, it remains the object of a social construction.

The second factor is related to memory. Papworth *et al.* (2009) distinguish between different types of baseline shifts: generational amnesia, where new generations are unaware of past biological conditions, and personal amnesia, where people lose memories of their own experiences. The authors also uncover the phenomena of illusory memory, where people relate a past state of the ecosystem that is not confirmed by scientists, and change blindness, where people continue to refer to the baseline without noticing changes that have affected the ecosystem. Alteration and distortion of memory will therefore influence the baseline.

The third factor involves experiences. Keilty *et al.* (2016) show how everyday life near dams can change their perception, to the point of being integrated into a landscape considered "normal" or even "natural". The role of experience is thus central, and this will have strong consequences. Thus, urbanization implies a loss of experience of nature, and thus the risk of a shifting baseline syndrome increased (Turner *et al.* 2004). The risk is then to take ordinary nature as a baseline, which is a considerably impoverished nature compared to wild nature (Maris 2018).

The fourth factor concerns cultural norms. In the Grands Causses, for example, it was from the 1970s onwards that the heritage and ecological value of open landscapes was progressively emphasized, whereas these landscapes had previously been considered as degraded landscapes, the baseline being that of a forest climax to be restored. This change in the baseline is closely linked to an evolution in the social representations of farmers. While pastoral societies were considered disrespectful to the natural equilibrium of the environment until the first half of the 20th century, they have gradually been recognized in their heritage role since the 1970s (Marty and Lepart 2001). The Cévennes National Park, which was initially conceived as a forestry park, thus made the maintenance of open landscapes the target of its management, which also made it possible to obtain the support of the local population, which was strongly agricultural (Basset 2009). Through this normative shift, the baseline does not refer to a natural landscape but to a cultural one, thus emphasizing the influence of cultural norms.

Finally, biological or evolutionary factors are also mentioned (Keilty 2016) insofar as aesthetic or environmental preferences can be explained by the evolution of humankind in contact with nature (Kaplan 1987; Wilson 2012).

Finally, these factors combine differently depending on the period, the social group, as well as the individuals. Different representations of the baseline thus coexist within a social group, as we have shown in the case of Mont Lozère (see Box 4.3).

As we have seen, the actors of Mont Lozère are opposed in their representations of the landscape dynamics in progress.

A first type of actor, the majority, considers that the "closing of the landscape" constitutes a loss, both ecologically and culturally. According to this representation, the baseline was situated in the middle of the 20th century, at the peak of the agricultural development of the territory. While the baseline generally refers to a nature untouched by anthropic impact, Mont Lozère is an emblematic example of the development and conservation of cultural landscapes. This can be explained by two main factors. First, cultural norms were progressively put in place from the 1970s onwards, supported by new knowledge, which allowed the recognition of the heritage value of agro-pastoral landscapes. Second, collective memory plays an important role, as actors evoke memories and anecdotes experienced by their ancestors.

A second type of actor, mainly public sector foresters and also inhabitants who have come to settle in this region for its wilderness character, sees the increase in forest cover as an enrichment of biodiversity. They insist on the anthropic character of open landscapes, and question the temporality of landscape standards. Indeed, the forest constitutes the "natural" landscape of Mont Lozère for them, whether they look to the past (Middle Ages, or even before) or project themselves in the long term of ecological dynamics. For this, they rely on knowledge of the biodiversity associated with forest landscapes, knowledge that is new but in full development, for example around the question of "old forests".

Finally, among the four factors that condition the baseline, we retain three on Mont Lozère: memory, cultural factors and knowledge (see Figure 4.2). Conversely, evolutionary factors, identified by Keilty *et al.* (2016), did not emerge in the interviews. Similarly, experience seems to enter into the picture rather little: the inertia of representations is strong, forest landscapes continue to have a negative connotation, even though the inhabitants have sometimes lived their entire lives in a predominantly forested landscape.

Box 4.3. *Mont Lozère and the case of cultural landscapes*

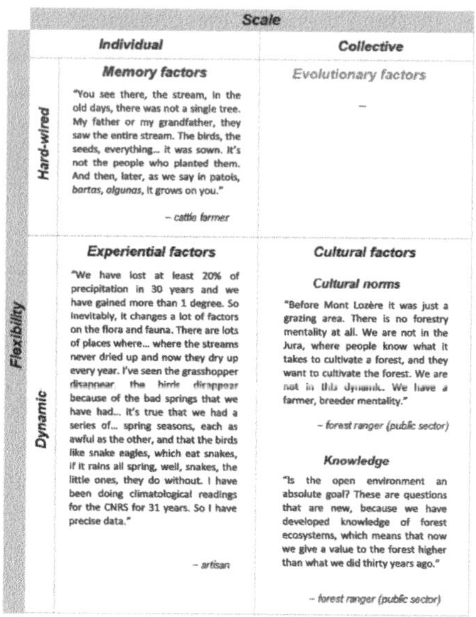

Figure 4.2. *Typical citations for the role of different factors identified by Keilty* et al. *(2016) in the construction of baselines on Mont Lozère*

4.4. Debating the baseline

The baseline has a descriptive value: it gives an image of the past state of the ecosystems, and makes it possible to give meaning to the dynamics in progress. But the baseline also has a normative value. Indeed, identifying a baseline is equivalent to choosing a precise point in the history of relations between societies and nature as a model. The baseline refers to the past, and also implicitly, to an "end state", an ideal or even a desirable future (Luginbühl and Terrasson 2013).

However, in the discourse of actors or in management documents, the baseline is rarely explained or justified, and often amalgamates an unrecognized and idealized historical state, persistent current states and a future ideal (Donadieu 2002; Davodeau and Barraud 2018). This vagueness surrounding the baseline makes it a unifying element in the governance of landscapes, since it serves as a medium for the expression of various, even incompatible, desires and fears (Davodeau and Barraud 2018). Thus, in mountain territories facing "landscape closure", the reference to agro-pastoral landscapes and the need to keep them open are brandished as a unifying element (Peyrache-Gadeau and Perron 2010), being the object of a consensus, at least in appearance (Barnaud and Couix 2020).

An illusion of consensus around the baseline then emerges, which can mask tensions between divergent representations, or make certain points of view inaudible. For example, in the Pyrenees, the consensus on the need to maintain livestock and open landscapes in reality conceals contradictory opinions, for example in favor of rewilding (Barnaud and Couix 2020). This illusion of consensus is sometimes supported by scientific arguments, presented as neutral, whereas knowledge is historically and socially situated, and the mobilization of knowledge by local actors is also done in a strategic perspective (Barnaud *et al.* 2021).

Finally, the baseline is neither scientific evidence nor necessarily a consensus, but rather a matter of choice. Indeed, the choice of a baseline will determine the feasibility and effort required to achieve conservation goals (Bull *et al.* 2014). It is therefore necessary to set up discussion arenas, in which the definition of the baseline and the definition of its role in management will be debated at the same time (Mathevet *et al.* 2015; Moreau 2019). The objective is then to recognize the legitimacy of each person to "make the landscape", by means of their practices or the enactment of standards (Hatzfled 2009).

Recognizing the plurality of viewpoints and their legitimacy within a constructivist approach is the basis of many participatory approaches (Barnaud and Mathevet 2015). The role of scientific knowledge in the definition of the baseline is then revised. Without falling into relativism, it is a question of recognizing the existence of other types of knowledge, such as local knowledge (Kai *et al.* 2014), and also recognizing that scientific knowledge is subject to numerous biases, and that scientists must also make the biases underlying their hypotheses explicit (Etienne 2010; Barnaud *et al.* 2021).

We have proposed to debate the baseline on Mont Lozère through a participatory approach to modeling (Etienne 2010) based on a role-playing game (see Box 4.4).

Our qualitative analysis revealed a significant divergence of views on Mont Lozère around the question "Open landscapes, yes, but which ones?" (Moreau *et al.* 2019a).

Debating the baseline through a participatory approach therefore seemed necessary, which we did as part of a SECOCO (Services ECOsystémiques et action COllective, Ecosystem Services and Collective Action, 2015-2019) research project, conducted in partnership with the National Park team. We designed a role-playing game, SECOLOZ, modeling the effects of agricultural practices on landscapes and the ecosystem services they provide. Two workshops were organized on Mont Lozère, with farmers, elected officials and National Park agents (Moreau *et al.* 2019b). The analysis of the exchanges during the parties and debriefing revealed two points of debate on the baseline.

First, participants debated the baseline's degree of naturalness. According to the farmers, the practices of rock removal and plowing of meadows are part of a long history of shaping landscapes through agricultural practices; there is nothing "natural" about these landscapes. For the National Park agents, the current changes are not of the same nature, due to their magnitude and irreversible nature. According to them, "naturalness" is not to be opposed to the "artificial" character of open landscapes: it is synonymous with biological richness. Thus, permanent or temporary meadows, although both resulting from human action, are in no way equivalent in their eyes in terms of biodiversity.

Second, during the game, the National Park agents drew the attention of the other players to the evolution of the game board, thus using the game to alert them to the risk of a shift in the baseline on Mont Lozère. On the contrary, the breeders consider that the risk of shifting from the baseline is low, in a conservation context that tends to sanctuarize landscapes. More generally, the farmers argue that the baseline can be considered as a starting point, but not as a management objective.

We thus proposed a concrete device for debating the baseline. The simplification of reality, the playful aspect and the exchange of roles allowed them to distance themselves from daily situations. The reflection initiated by the National Park, and supported by the SECOLOZ role-playing-game, led to a modification of the authorization system, in order to promote mutual understanding between the National Park and its citizens.

Box 4.4. *SECOLOZ, a role-playing game for debating the baseline*

4.5. Conclusion

Questioning the baseline leads us to collectively define a desirable nature and a desirable future, which are central in a context of global change. This review of the literature on the baseline and its shift, as well as our study of the changing landscapes of Mont Lozère, show the need to support the reflection of territorial actors with critical and integrated approaches, combining the study of social representations and practices, and the study of the ecology of the environments concerned. The material and immaterial dimensions of the baseline must be grasped both in the long term nature of their historical and political trajectory, in the plurality of points of view and the diversity of scholarly and lay knowledge of the evolutionary processes. The ecological, economic, social and institutional transition of the territories is always in search of new development models, values and projects for the transformation of the socio-ecological community. Will a better knowledge of the baselines (favoring a fine knowledge of our interdependencies through dialogue and action) be able to found new trajectories for our territories of life, and thus contribute to the well-being of humans and non-humans?

4.6. Acknowledgments

We would like to thank all the people interviewed during the work on Mont Lozère, the participants in the role-playing workshops and the SECOCO project team. This work was financed by the Ecoserv Metaprogram, the SAD department of INRAE and the Cévennes National Park.

PART 2

Using Baselines to Conserve Nature

5

Rewilding by the Return of Ghosts of the Past

5.1. Preamble

Seldom, Bonnie, George and Doc Sarvis[1] are all avid nature lovers, friends since college (although George was only there for three months). After blowing up a useless dam built a year ago and tearing down the huge billboards on Highway 60, the four friends were not unhappy to leave Phoenix and go camping near Burro Creek. But Seldom suddenly stops their car. A plume of dust rises in the distance, signaling the advance of a group of fifteen mustangs, proud little horses with colorful coats, barely four feet tall. Seldom is fascinated by these animals. The three others know that he is going to give them a lyrical flight of fancy.

"You see folks," Seldom says as expected, "here's what's left of our good old American nature, just as it was in its origins, just before the Westerners screwed it up!"

"Are you kidding, Seldom?" Bonnie shouted out. "These animals don't belong here. We're really into the cliché of cowboys and Indians. It's a real symbol of the introduction of the Europeans and their domination of a whole nature that was still untouched when they arrived! It is the very image of the domination of the white man for me."

"I think we should protect and even enhance the populations of these animals because there are no more of them in much of the state!" said George. "In college, in the second week of class, our professor talked about 'keystone species' for these kinds of big beasts. If we want to get back to ecosystems with some semblance of wilderness, our four-legged gardeners are right there in front of us!"

Chapter written by Laurent GODET.

1 Character names borrowed from Edward Abbey's *The Monkey Wrench Gang.*

Doc Sarvis, who had been silent from the start, and was not known to be a great talker, scratches his throat and sighs:

> "Shall I let you two fight it out or shall I remind you, Seldom and Bonnie, that you are both wrong and that George is just speculating like a lunatic, which is what he is? I disagree with you Seldom: there were no horses when the Spanish landed in America. But I also disagree with you Bonnie … because if the Europeans did bring horses to this continent from which they were absent at the very moment of their arrival, there were horses long before them. Horses were present until the end of the Pleistocene in North America, but were eradicated."

George, believing that he has finally found favor with Doc Sarvis, exults:

> "I'm right, we have to protect these animals! We even need to reintroduce more of them! Everywhere."
>
> "It depends on what you want to find, old George!" answered Doc Sarvis. "You're mixing everything up in the same sentence. Species are not just agents that "do something" in an ecosystem. As an evolutionist, I remind you that species not only *do* something, they *have* something. Morphology, genetics, etc. Pleistocene horses are very different from those of today. Good old George, nature is not interchangeable. Secondly, you are making a perhaps risky assumption that you are going to get back to the ecosystems of yesteryear by reintroducing horses. What exactly is known about the structuring role of these large herbivores? And will they find ideal living conditions to thrive in a climate and on soils that are very different from tens of thousands of years ago? And, finally, and I'll stop here because you are already asleep: reintroducing horses on the grounds that they were Pleistocene fauna, *i.e.*, animals that were present before the great extinctions of the megafauna due to humans, is perhaps arbitrary, isn't it? At this rate, we must also reintroduce mastodons, big cats, camelids, etc. into North America. In short, let us start a life-size Jurassic Park?"

Doc Sarvis was always incisive. But, more than enlightening his audience, he always plunged them into an abyss of questions that seemed insoluble. If his three companions always wanted to act vis-à-vis the environmental degradations, but in directions often quite different, Sarvis always cut them down by saying that they did not know enough. Quoting Paul Valéry without knowing it, he sighed all day long:

"How many things one must ignore to act!" He kept on saying that before (re)introducing a species, it was necessary to know where it had come from, why it had disappeared, what its functional role was and according to what practical, ethical and even political modalities. But this evening, he no longer had the strength to give them a lecture. This chapter presents what Sarvis would have wanted to develop if his three sidekicks had not already left to light a campfire and open beers by the edge of the river.

5.2. Introduction

Scientific studies and reports from different NGOs are now accumulating to show us the magnitude of the sixth biodiversity crisis (IPBES 2019) that affects a large number of the taxonomic groups studied (Barnosky *et al.* 2011; Ceballos *et al.* 2015).

Part of the work of conservation scientists is to record and monitor this crisis, but as a discipline of crisis and urgency (Soulé 1986), it also proposes solutions to counter it. While the cornerstone of proposed measures is the creation of protected areas and the implementation of integrated management measures, a significant portion of scientific studies in recent years also propose restoration measures in a broad sense, focused on species as well as habitats (Godet and Devictor 2018).

Schematically, we can therefore consider that some of the conservation measures seek to slow down the biodiversity crisis by establishing bulwarks in the face of the threats to biodiversity, but that another part, quite complementary, clearly seeks to "repair" it. The erosion of biodiversity should not therefore be seen as a steamroller that is inexorably advancing and whose course we are simply trying to slow down. Restoration measures aim to precisely return to before the tipping points that have led to the disappearance of habitats or species. In Western Europe, a very large number of the protected areas have obviously opted for a return to an agro-pastoral nature, which corresponds roughly to the conservation of rural landscapes before the diffusion of the second agricultural revolution (arising at the end of the 19th century, but especially diffused in the aftermath of World War II with mechanization and the generalized use of chemical inputs). It is a question of maintaining both cultural landscapes and the biodiversity associated with them (see Chapter 11). Here, we find the classic ingredients of sustainable development with a "harmonious integration" of the social, the economic and the environmental (see Agnoletti (2014)). But a very different form of "turning back the clock" is also proposed, mainly in environments where agro-pastoral landscapes have been the object of less patrimonialization: this is "rewilding".

Rewilding consists of restoring a wild dimension to nature, *i.e.*, loosening the human stranglehold on natural dynamics by restoring a character and functioning similar to what it may have had before anthropic domination. In contrast to the conservation of agro-pastoral landscapes, we can, therefore, say that this movement seeks to move the temporal cursor of the functioning of ecosystems and also of the desired reference state a little further back in time. We can distinguish two currents in rewilding strategies: a rather passive form and an active one. The passive form consists, above all, of giving freedom to ecological processes: human activities are withdrawn from a space or we take advantage of their withdrawal (in the case of agricultural abandonment, for example, or human depopulation of certain spaces, etc.). It is here that the spontaneity of processes and the feralization of environments are sought (see Navarro and Pereira (2012); Schnitzler (2014)). We will not address this aspect of rewilding here because a chapter in this book is devoted to it (see Chapter 6). The active form consists of restoring the autonomous functioning of environments in order to recover ecosystems similar to those that have disappeared by (re)introducing the so-called "keystone" species that will then reshape ecosystems and allow them to function without the need for human intervention. Even though we are aware that this division is somewhat arbitrary, since several rewilding projects link the two approaches (*e.g.*, the sites for the reintroduction of keystone species are often abandoned areas that are the subject of land acquisition for conservation purposes), we will focus solely on "trophic rewilding", which corresponds to the (re)introduction of species at the top of food chains, in order to re-launch natural dynamics and recover disappeared ecosystems.

In this chapter, we will first look at the large carnivores or herbivores that have disappeared and try to understand what were the causes and consequences of their disappearance and extinction. Second, we will explore ways of rewilding through different schools of thought and several concrete examples. Finally, this chapter will conclude with the criticisms and controversies raised by rewilding.

5.3. Contemporary ecosystems populated by ghosts?

5.3.1. *Extinctions and disappearances of species*

Planet Earth is 4.53 billion years old, and it is fairly commonly accepted that life appeared on it 3.5 or 3.6 billion years ago, although recent (and still hotly debated) studies argue for an even earlier origin at 4.10 (Bella *et al.* 2015) or even 4.28 billion years ago (Dodd *et al.* 2017). Since the appearance of these first life forms, it is estimated that about four billion species have appeared on Earth, and 99% of them have become extinct at some point in this long history (Novacek 2001). The extinction[2] of species is therefore a common phenomenon on Earth, which is

2 In this chapter, we will distinguish between "extinctions" and "disappearances": the former corresponding to the total disappearance of a species from the planet (such as the Dodo,

balanced with the processes of speciation (*i.e.*, the appearance of new species). However, this history of life, during which we can calculate an average extinction rate per period of time, has been punctuated by five major crises. These "extinction crises" are generally accepted within the paleontological community as brief episodes on the scale of geological time (a few million years or less), during which three quarters or more of species become extinct on a global scale (see Barnosky *et al.* (2011)). The first crisis was the Ordovician (443 million years ago, with 86% of species extinct), followed by the Devonian (359 million years ago, 75% of species extinct), the Permian (251 million years ago, 96% of species extinct), the Triassic (200 million years ago, 80% of species extinct) and the Cretaceous (65 million years ago, 76% of species extinct) – for a synthesis of these five crises and the associated bibliographic references see Table no. 1 in the article by Barnosky *et al.* (2011).

We are currently experiencing a sixth extinction crisis of life that is affecting many taxonomic groups and is characterized by very large extinction rates in recent centuries (Leakey and Lewin 1996). The history of this crisis is intimately linked to that of the expansion of modern humans *Homo sapiens*; so much so that some authors refer to it as an "unnatural history" (Kolbert 2014). Let us explore the different time steps at which this crisis can be apprehended, the species affected and the spatial extent of this phenomenon.

A first scientific corpus has largely documented the extinction of a large number of the "megafauna" (*i.e.*, animals weighing more than 44 kg as adults – Martin (1984)) at the end of the Pleistocene. As Paul S. Martin[3] reminded us in some of his lectures, we are living in a land of ghosts since all of the major species that inhabited our planet a few thousand years ago have disappeared. The fascination of researchers with the great extinct species is quite old. In the 19th century, in the Origin of Species, Charles Darwin (1859) reported the discoveries of Messrs. Lund and Clausen in Brazil and of Professor Owen who described giant extinct birds in New Zealand, and also his own observations about the remains of megatheriums (giant sloths), mastodons, toxodons and "other giant mammals" in South America. His contemporary Alfred Russel Wallace wrote that we clearly live in a zoologically impoverished world because it is devoid of the most "immense, ferocious, and strange" mammals that have all suddenly disappeared (Wallace (1911) cited in Elias and Schreve (2007)). The 19th and 20th centuries then helped to date and locate these extinctions precisely. The Pleistocene megafauna mainly concerns large

endemic to Mauritius, which has disappeared on a global scale), and the latter corresponding to the local or temporary disappearance of a species (such as the Monk seal *Monachus monachus* in France, which disappeared from Corsica in the 1970s, but of which a few populations are still present in Madeira, northwest Africa, or in the eastern Mediterranean and the Black Sea – Pascal *et al.* (2006)).

3 Geoscientist and Pleistocene extinction specialist at the University of Arizona.

herbivorous mammals such as mammoths and woolly rhinoceroses, and also some giant flightless birds, turtles and giant lizards in Oceania (Elias and Schreve 2007). This extinction is indeed global, although some continents such as Australia and the Americas have been more affected than Asia and Africa, or even Europe. The causes generally put forward in the literature concern the older human settlements on these last three continents, which allowed the large species to better adapt to and coexist with humans (Martin 1984). This crisis is spread over several thousands of years and the dating of these extinctions can still evolve with the dating of new fossils. Even though these extinctions were attributed to human action very early on (as early as 1911 by Wallace – see Elias and Schreve (2007)), there is a debate that still rages on the cause of these extinctions since the 1960s, between supporters of the anthropic origin (see the "overkill hypothesis" by Martin (1967, 1984)) and defenders of the environmental hypothesis – mainly climatic (or at least the detractors of the anthropic hypothesis: see Grayson and Meltzer (2003) regarding the case of North America). Elias and Schreve (2007) and Stuart (2015) have drawn up reviews of the arguments of the two "camps", but specify that the debate will remain sterile in the absence of new precise dating of the remains of these species. However, Bartlett *et al.* (2015) have established a database of the most recent dates of presence of each species in parallel with a reconstruction of climate change and human colonization, and conclude that human colonization is the main factor in the extinction of terrestrial megafauna, even though climatic factors may have also played an important role. Let us also recall two alternative hypotheses. The first one puts forward the indirect role of humans on these extinctions by the contribution of diseases, the vectors being humans themselves, as well as their commensal species (MacPhee and Marx 1997). The second is an intermediate hypothesis stating that the extinctions are due to climatic factors and that humans would have dealt a fatal "coup de grâce" to the megafauna through their hunting activity (see Elias and Schreve (2007)).

A second body of literature comes from archaeozoology. On the scale of the Holocene as a whole (last 10,000 years), we note an acceleration of local extinctions and disappearances of vertebrates, which affected mainly birds and mammals. Taking France as an example, Pascal *et al.* (2006) note a secular index of vertebrate extinction (number of species extinct or disappeared from France per century) of the order of 0.2 at the beginning of the Holocene, 1.5 over the period 1600–1800, 9.6 over the period 1800–1914 and more than 12 species extinct per century from 1914 to 2004. These extinctions affected 56% of birds and 36% of mammals (Pascal *et al.* (2006) deal "only" with vertebrates in their book). On the scale of the end of the Holocene, *i.e.*, during the last thousands of years, it is indisputable that a large number of the extinctions of small and large vertebrates on oceanic islands are due to humans (Alcover *et al.* 1998; Martin and Steadman 1999; Turvey 2009; Crowley

2010). Jean-Denis Vigne[4] has devoted a great deal of work and given numerous lectures on the extent of extinction of endemic vertebrates and the introduction of new species in Mediterranean islands due to human activity (see, for example, Vigne (1988, 1992); Vigne and Valladas (1996) for Corsica and Sardinia, or Vigne (2004)). Corsica was only populated by humans later on, in the Mesolithic (around the 8th and 7th millennia BCE), by mobile hunter-fishermen. The first Neolithic villages on the island date from the 5th millennium BCE. Among the six species of Corsican mammals present in the Tardiglacial period, the extinction of two of them around 10 to 9,000 BCE is attributed to a sudden climatic warming: the Sardinian Dhole (*Cynotherium sardous*), an endemic canid of Corsica and Sardinia (although Gippoliti and Amori (2006) now consider its extinction as intervening shortly after the arrival of humans on the island) and the Caziot's Deer (*Megaloceroides cazioti*). The other four species: the Sardinian pika (*Prolagus sardus*) (small rabbit with short ears), the endemic Corso-Sardinian Mulot (*Rhagamys orthodon*), the Corso-Sardinian endemic vole (*Tyrrhenicola henseli*) and the Corsican endemic shrew (*Episoriculus corsicanus*) (all of a large size) have resisted the climatic shock and persisted throughout the Mesolithic (period during which the pikas were intensively hunted by humans) and the Neolithic (period during which these species probably suffered a lot from predation by foxes and dogs, both introduced by humans). These species became extinct later, in the first millennium, notably as a result of major deforestation and the development of agropastoralism in late antiquity (Pascal *et al.* 2006). Compared to these extinctions, several of which are directly attributable to the arrival of humans on the island, it can be said that all of the other mammal species on the island, 26 species, were introduced by humans and were therefore not present before the arrival of the first Neolithic populations (Vigne 2004). They are, in order of appearance on the island, the following species: sheep, goat, pig, fox, dog, ox, mouflon, wild boar, hedgehog, field mouse, European edible dormouse, mouse, shrew, Etruscan shrew, garden dormouse, horse, donkey, black rat, cat, wildcat, deer, hare, weasel, bear, brown rat and rabbit (Vigne 2004). The island's mammalian fauna has thus been completely modified, with its original six species replaced by 26 new ones. This increase in alpha diversity (island scale) was accompanied by a sharp drop in gamma diversity (Mediterranean basin scale) because most of the species that disappeared were endemic to the island or to the group of neighboring islands. This Corsican example, with the extinction of an endemic mammalian population poor in species, followed by a massive introduction of numerous species by humans, can be found on several Mediterranean islands: Sardinia, Majorca, Crete and Cyprus (Vigne 2004).

On shorter time scales, the International Union for Conservation of Nature (IUCN) has chosen the year 1500 as the arbitrary date from which species

4 Director of research at the CNRS, archaeozoologist at the Muséum National d'Histoire Naturelle de Paris.

extinctions can be considered "recent". To date, the IUCN has recorded 108 species of mammals, 141 species and 138 subspecies of birds that have become extinct since 1500 across the globe. Szabo *et al.* (2012) have compiled an inventory and map of where birds have become extinct since 1500. The overwhelming majority of extinctions are in island environments, coastlines and the intertropical band. All extinctions are attributable to humans, except for the Rock Wren subspecies *Salpinctes obsoletus exsul*, which disappeared following a volcanic eruption on the island of Revillagigedo in Mexico (Brattstrom 1990). The main causes of extinction are, in descending order: introduction of exotic species (58.2% of species; 50.7% of subspecies), hunting (52.4%; 18.8%) and agriculture (14.9%; 31.9%). Ceballos *et al.* (2015) thus estimated an extinction rate over the last 100 years that was 100 times greater than the average species extinction rate for vertebrates.

Since the 1980s, conservation biology has placed much emphasis on the conservation of large, rare and charismatic species, rightly considered as threatened (Soulé and Wilcox 1980). In France, Barbault (1997) assumed that the sixth extinction crisis (which he described as the Holocene crisis) affected birds and mammals in particular. But gradually, it became clear that many taxonomic groups were also impacted, such as amphibians (McCallum 2007) or insects (Fonseca 2009; van Strien *et al.* 2019). More than species richness, we are also dealing with a drop in the abundance of common species. Participatory ecology programs, which have enabled the monitoring of common species, have, for example, shown significant declines in common country birds (Donald *et al.* 2001). Recently, several NGO publications have warned the public about the collapse of animal abundances globally, with a 68% average decline in species population sizes since 1970 (WWF 2020).

In summary, it is safe to say that we are in what can be called an extinction crisis, because the current erosion of biodiversity: (1) is recent and rapid; (2) is leading to the disappearance of a large proportion of taxa; (3) is a planet-wide phenomenon. To appreciate the rapidity of the current extinction crisis, let us relate it to a distance. If the history of life on Earth were related to a trip between Moscow and Paris that a crow flies (2,486 km), the Ordovician crisis would appear at kilometer 2,222, *i.e.*, 263 km from Paris, the Devonian crisis at 222 km, the Permian crisis at 150 km, the Triassic crisis at 121 km and the dinosaurs would disappear in the Cretaceous at only 39 km from Paris. Finally, *Homo sapiens* would only appear less than 200 meters from Paris and the great extinctions of the end of the Upper Pleistocene would make their appearance on the Paris ring road, less than 10 meters from the capital! Whether the sixth extinction crisis began at the end of the Pleistocene, in Antiquity, in the Middle Ages, in 1500 or 1950 is of little importance on the geological time scale, since this crisis is so recent. In such a short time, we can nevertheless see a progression of the crisis since the extinction of the very large species of the megafauna (Upper Pleistocene), of mammals and birds throughout the

Holocene, with a marked extinction of mammals and birds endemic to the island environments progressively colonized by humans (from Antiquity), then what remains of large herbivores (such as the Aurochs and the Horse in Eurasia) and large predators during the modern period, and finally, since the last decades, a continuation of the crisis that extends to a large part of the taxonomic groups and affects their abundances this time. In a schematic way, disappearances and extinctions therefore affect smaller and smaller species (as demonstrated by Dirzo *et al.* (2014)), belonging to more and more varied taxonomic groups. But beyond the disappearance and extinction of these species themselves, what may have been the consequences of these disappearances and extinctions on ecosystems and their functioning?

5.3.2. *Impacts of extinctions and disappearances on ecosystems*

Substantial literature has focused on the consequences of the disappearance/ extinction of large herbivore and carnivore species on ecosystems.

The impacts of Pleistocene megafaunal extinction on ecosystems are quite difficult to assess. Most studies rely on the impacts of contemporary equivalents to infer the functional role of extinct species. In a review of the potential impacts of tropical megafauna extinction, Corlett (2013) lists six potential impacts: (1) co-extinction of parasites; (2) co-extinction of commensal and mutualistic species; (3) changes in abundance of competitors (increase in small herbivores following extinction of large ones – but this is poorly documented except in Madagascar where species under 10 kg appear to have increased greatly after late Holocene extinctions (Crowley 2010)); increase in meso-predators after extinction of apex predators (see Sutherland *et al.* (2011); Ritchie *et al.* (2012)) – but this is still poorly documented from a paleontological perspective; (4) impacts on prey–predator relationships and scavengers: Pleistocene equivalents such as wolves (Hofreiter and Barnes 2010) or coyotes (Meachen and Samuels 2012) were often larger than contemporary species due to larger prey and competitors; (5) impacts on vegetation with the maintenance of a mosaic of habitats as observed today through African elephant grazing; (6) impacts on methane, although this is likely insufficient to explain climate change (Shakun *et al.* 2012). For Owen-Smith (1987), the extinction of mega-herbivores (which he attributes to human hunting) led to very large habitat changes from a mosaic of heterogeneous habitats to more uniform grassland or forest habitats, which, in turn, were detrimental to small herbivores that depended on the mosaic, nutrient-rich habitats maintained by the mega-herbivores and which then became more vulnerable to human predation.

For the prehistoric megafauna (early Holocene, spanning from about 9,000 to 5,000 years BP), there is a debate not only about what the "original" environments of the early Holocene were, but also about the impacts of large herbivores on forest cover. There are two main schools of thought. On the basis of palynological analyses in peat and lakes, the first, which we will call the "closed forest" hypothesis, assumes that the European plains (outside of high latitudes) were covered by a closed mixed deciduous forest until 6,000 years BP and the appearance of human impacts (Firbas (1949) and Iversen (1960, 1973) cited in Birks (2005)). In this hypothesis, the abundance of large herbivores was controlled by forest structure. Vera (2000) proposed a different view. For him, during the early postglacial period, the European plains were made up of a mosaic of open grasslands, scrub and forests. This is the "wood-pasture hypothesis" (proposed by Vera (2000)). Vera (2000) relies on pollen analyses in bogs and lakes showing the past abundance of oak and hazelnut trees, which require canopy gaps to regenerate. According to him, these gaps were maintained by the grazing of large herbivores (*e.g.*, bison, aurochs, horses, red deer). Grazing by herbivores in forested areas prevented their regeneration, opening up the canopy and turning the forest into a "park". The growth of scrub in this "park" then provided exclusion zones for herbivores, allowing young trees to develop and then form new groves. It can be noted that the pattern proposed by Vera is quite close to fairly old visions in forest ecology, in particular the "gap-phase dynamics" of Watt (1947), according to which the mortality of old trees allows plant successions to take place; the result is a mosaic forest with its different stages of plant succession. The main difference is that according to Watt (1947) and his successors (*e.g.*, Larsen 1980; Heinselman 1981) the driving force behind these successions is the mortality of old trees by various disturbances such as fire or insect pests. However, the "wood pasture" hypothesis as an original environment of the European plains has been overturned by the work of Mitchell (2005), who demonstrated that the landscapes of this period were indeed dominated by a closed forest. Mitchell (2005) first relied on pollen analyses from Ireland that he compared to continental Europe. Since during the post-glacial period, there were no or almost no large herbivores (except for Red Deer and Wild Boar in low abundance). Ireland was a control site that was not subject to grazing. Following Vera's hypothesis, there should be no open areas since there were no large herbivores to create them. However, in the pollen analyses of the period 6,500–9,500 years BP of the peat and lakes of Ireland, Mitchell (2005) found just as many oaks and hazel trees as in England and continental Europe, indicators of open canopies, which makes it possible to say that the large herbivores did not control the degree of opening of the forest. In a second step, he compared pollen analyses within forest depressions from six sites in Sweden, two in Denmark and four in Ireland. The contemporaneous pollen analyses show that the presence of 60% or more tree pollen is an indicator of closed forest, while the presence of 50% or less tree pollen indicates open environments. Mitchell used these ratios to reconstruct the presence of closed forest or open areas at sites sampled in Sweden, Denmark and Ireland. He found that the presence of tree pollen only dropped below 60% in the last 3,000 years, which corresponds to the beginning of anthropogenic impacts. If we follow Mitchell's results, the original postglacial environments of the European plain were closed forests. The openings in these forests are more recent, dating back 3,000 years, and are due to human action. It should be remembered, however, that both of these hypotheses are based on rather ad hoc pollen analyses and that it is assumed that the ecological requirements of the species were the same as today (see Birks and Birks (1980)): in other words, oak and hazel trees were not necessarily indicative of open environments several thousand years ago and the ecological niche of herbivores may not have been the same either. There is also a significant knowledge gap regarding the *densities* of large herbivores in the past (see Bradshaw *et al.* (2003)), making it impossible to know their real impact on forest structure. Even though we follow the reasoning of Mitchell (2005), this does not negate the potentially important role that large herbivores can play on forest structure. It is known, for example, that changes in abundance and space used by large herbivores in the presence or absence of large predators can be of fundamental importance to habitat dynamics.

For the recent period, several studies inform us about the functional role of apex predators. The loss of large predators is not only an ethical or aesthetic issue for Estes *et al.* (2011), but also has had far-reaching ecological consequences that these authors refer to as "trophic downgrading". Three key concepts in ecology are necessary to understand what this is all about: (1) ecosystems are shaped by apex predators (see Crooks and Soulé (1999)) and there are "trophic cascades": when one apex predator disappears, all lower trophic levels are affected (see Paine (1980)), (2) ecosystems should be seen as networks of complex interactions among all species and in all directions (see Estes *et al.* (2011)) and (3) there are "stable alternative states" of ecosystems: these are not always stable, but can change dramatically with disturbances to regain a new stability (see Kéfi (2012) for a very informed popularization of these concepts). These three concepts allow us to understand the principle of trophic degradation: the disappearance of a apex predator simplifies the trophic chain, modifies the pressure exerted by herbivores and thus the composition and abundance of plant species, and these changes can sometimes be rapid and not very reversible.

Several examples of the consequences of the disappearance of predators on ecosystems have been documented, such as the sea otter (Estes and Duggins 1995) or the polar fox (Croll *et al.* 2005). But the reintroduction of the Grey Wolf in Yellowstone Park in the United States gives us one of the best-documented examples of the ecological change induced by its presence. We can therefore imagine, in reverse, the consequences of the extinction of large predators in the past. The wolf disappeared from Yellowstone in the 1920s and was reintroduced in 1995–1996. Fifteen years later, Ripple and Beschta (2012) note a trophic cascade effect on ecosystems: wolves regulated elk populations (they decreased their abundance and restricted them to certain areas), brush and woody vegetation increased in some areas (especially along streams that elk frequented less), and

Canada beavers and American bison increased, probably because of the greater abundance of woody and herbaceous vegetation due to the decrease of elk. In particular, the recolonization of beavers has altered the waterways. The consequences, in only 15 years, of the reintroduction of the wolf have been spectacular, modifying the abundance and distribution of other species, and even changing the physiognomy of the waterways.

To conclude this part, we can therefore say that we are in a sixth extinction crisis that began recently (at the latest at the end of the Pleistocene), which first affected the largest species belonging to the megafauna at the end of the Ice Age, then mammals and birds throughout the Holocene (with a particular impact in island environments), and which today affects the abundances of the most common species (notably common birds) of increasingly modest size (notably insects). The origin of this crisis, contrary to the five previous ones, is of anthropic origin, even though for the most remote times the question of the part of the influence of humankind as "chief extinguisher" of the big species or as simple carrier of a "coup de grâce" still remains open. Ecological works inform us about the consequences of these multiple extinctions. Even though it is very difficult to know the exact impact of the disappearance or extinction of large fauna on ecosystems over the last thousands of years, the impacts of functionally equivalent contemporary species show us that when a large predator or herbivore species disappears, ecosystems can rapidly change and sometimes in a way that is difficult to reverse.

Since humans have led, at least in part, to a progressive "defaunation" (Dirzo *et al.* 2014) of the Planet over the past several millennia, and this has had – and is still having – significant and near-irreversible consequences on ecosystems until these species return, the resettlement of large extinct or extirpated species, which is a form of rewilding, is then proposed as a solution to turn things around.

5.4. Rewilding to repair

5.4.1. *How far back do we go to rewild?*

Rewilding is a form of ecological restoration, itself understood as "the process of assisting the recovery of an ecosystem that has been degraded, damaged, or destroyed" (Society for Ecological Restoration Science and Policy Working Group 2002). Ecological restoration can take a multitude of forms (see Higgs (1997)) depending on the objectives to be achieved (what baseline is desired, is the aim more anthropocentric or ecocentric, etc.) and the methods to be used (in particular, how much of it involves the maintenance or restoration of human activities). In the context of rewilding, restoration aims, in fact, to restore a "wild" character, *i.e.*, one that is distinct from the "domestic" and therefore from domination by humans. As recalled in the introduction, rewilding can take a first form, which is the free evolution of environments, either by taking advantage of the abandonment of human activities that already exists (most often) or by encouraging it (notably through land control). This first form of rewilding is not dealt with in this section because it is the subject of an entire chapter in this book (see Chapter 6). The second form of rewilding, referred to as trophic rewilding, is the re-introduction of large herbivore and predator species to recover lost ecosystems and their functioning.

The main question raised by any form of rewilding through the resettlement of large wildlife is: what is the desired baseline? As Callicott (2002) points out in relation to ecological restoration projects, the question seems fairly obvious at first glance: we need to go back to "original" conditions, which can be understood as a state prior to degradation by humans.

A first body of literature therefore logically explored the idea of a restoration of Pleistocene fauna and, consequently, of associated environments. As we will see later, this "rewilding of the Pleistocene" has been developed both in the United States and in Russia. It is based on two observations: it is necessary to return to a natural state and these natural conditions correspond to a quasi-non-anthropic influence. If the end of the Pleistocene and the beginning of the Holocene corresponds to Paul S. Martin's "overkill" seen earlier in this chapter, then we must return to just before that tipping point. One of the first to imply the need to return to this Pleistocene fauna is Soulé (1990):

> For many North American ecologists, the psychological adjustment to biogeographically recombined communities will be painful, but it may be facilitated by the realization that lions, cheetah like cats, camels, elephants, horses, saiga antelope, yaks and spectacled bears are native taxa to North America that disappeared very recently. The reintroduction of these large animals will be controversial, but I would not be surprised to read someday that cheetahs are helping to control deer and that mesquite is being "overbrowsed" by rhinoceroses. A cheerful way of viewing such faunal mixing is that it represents the restoration to the Nearctic of the great paleomammalian megafauna.

But the most fervent advocate is Paul S. Martin (see Martin and Burney (1999)), who is one of the first to clearly propose the establishment of Pleistocene parks.

For Callicott (2002), the appropriate baseline for restoration (and thus rewilding) must be the ecological (not evolutionary) time scale. Callicott draws on Holling (1992) to distinguish three scales: microscale (diurnal, seasonal, interannual cycles), mesoscale (environmental and disturbance processes) and macroscale (geomorphic and evolutionary processes). For Callicott, the microscale is too small and the macroscale is too large to deal with restoration. Humans have life cycles that they generally divide into 5–10 year increments (people talk about their childhood, adolescence, thirties, forties, etc.), which correspond roughly to microscale cycles such as vegetation cycles, whereas the dynamics of the processes (and disturbances) are on a mesoscale, *i.e.*, on time steps of centuries. For Callicott, the Holocene corresponding to a few hundred centuries would be an ideal scale to practice restoration. At these time scales, the Pleistocene extinctions have passed, humankind is already an apex predator, and the species that remain are the survivors of what he calls "the ecological holocaust". But Callicott further restricts this Holocene window to the very end of the Holocene when climatic and soil conditions were similar to those of today.

In more recent times, some authors have also proposed a rewilding in a pre-colonial era for North America. But this idealization of a state before human settlement is based on two myths: that of the wilderness (see Cronon (1996)), which would be an intact nature before the arrival of settlers; and that of Clements' climax (Callicott 2002). For Clements (1916), ecosystems are a kind of hyperorganism determined by the climate (mainly temperature and precipitation) and there are climactic plant formations for each biome. According to Clements (1916), humankind is a disruptive element of this functioning. But this vision was contradicted in the first part of the 20th century, during which we moved from an idea of the balance of nature to the "flux-of-nature": there is no homeostatic character of ecosystems and trying to define (and a fortiori restore) "original" states of communities or ecosystems is illusory. We know today that the idea of equilibrium states of ecosystems is challenged by empirical and theoretical studies that highlight the following points (Pickett 2013): ecosystems are open, they can be regulated by the outside, the absence or temporary nature of equilibrium states, the importance of humans in the functioning of ecosystems and the disturbances they generate, and the multiplicity of potential pathways of ecosystem dynamics.

Finally, many ecological restoration programs advocate a return to (often idealized (Agnoletti 2014)) cultural landscapes. These cultural landscapes can be understood as a landscape *fashioned from a natural landscape by a culture group. Culture is the agent, the natural area is the medium, and the cultural landscape is the result* (Sauer 1925). This is outside the scope of "rewilding" per se because these projects do not involve the relocation of large *wild* carnivores or herbivores, but rather the introduction of domestic animals (the archetypal example being the introduction of Highland Cattle, which are reputedly hardy and "wild" only in appearance).

5.4.2. *How is rewilding achieved? American, Russian and European trends*

It is interesting to note that rewilding projects and their conceptualization appeared in different parts of the world, namely, in North America, Europe and Russia in the 1980s–1990s.

In North America, Soulé and Noss (1998) proposed rewilding as an inseparable complement to the classical biodiversity conservation of the time – understood as a network of protected areas encompassing a representative diversity of ecosystems. Almost as a slogan, they define rewilding as the three Cs rule: Cores, Corridors and Carnivores: *i.e.*, large cores of strictly protected nature (the "wild"), ecological corridors linking these cores, and keystone species composed of large predators (although the authors note that other species can also be considered "structuring"). Their roadmap is clear: restore large predator populations and allow a trophic cascade to develop. Large predators play two major roles in conservation: maintaining the "integrity" of ecosystems and acting as an "umbrella", since their protection requires the protection of very large areas that must also be connected to each other. Soulé and Noss believe that rewilding will allow the "self-maintenance" of ecosystems. Finally, they argue that even though rewilding may seem, at worst, like an atavistic obsession with the resurrection of an Eden or, at the very least, like a somewhat utopian vision, it is nevertheless a matter of scientific realism. Soulé and Noss justify this realism by the *means* of rewilding: keystone species have a vital role to play in maintaining the ecological structure, diversity and resilience of nature, and also by the purpose of rewilding: according to them, it is a moral obligation to protect the wilderness and keep the "survivors of the Pleistocene" alive not only for the pleasure of humans, but also for their intrinsic value. In short, the rewilding proposed by Soulé and Noss is a continuation of Leopoldian thinking, in which the integrity, stability and beauty of nature were to be preserved (Leopold 1949). Soulé and Noss speak of the integrity of ecosystems, and, as informed ecologists, prefer to speak of "self-maintaining" ecosystems rather than of ecosystem stability per se, since this notion, like the idea of *climax*, has been debated among ecologists. As for beauty, even though they deny it (they would not have the will to restore an Eden and would not only be utopians), the restoration of the great fauna founding the identity of a whole continent before its anthropization is indeed something that comes from the admiration of the beautiful and the great (even though the authors admit half-heartedly that other species are structuring, like corals, etc., they only choose to speak about the great species). The thinking of Soulé and Noss continues with authors like Josh Donlan who propose a rewilding of North America in a very concrete way, without detours or language. Donlan *et al.* (2005) published an article that would become a landmark in the world of conservation: they proposed the reintroduction of functional equivalents to the large predators and herbivores that disappeared during the Pleistocene. For these authors, the baseline of the arrival of Europeans in 1492 is not acceptable and it seems less arbitrary to them to fix this to -13,000 years, the date of arrival of the first humans from Eurasia in North America via Beringia. Moreover, they argue that African and Asian elephants can be introduced as a substitute for Mastodons, Mammoths and Gomphotheres, the Bactrian Camel as a substitute for the extinct American Camelids, the African Cheetah as a substitute for the American Cheetah (or Miracinonyx), etc.

Zimov *et al.* (1995), in a seminal article in The American Naturalist, review the predominant role of hunting as a factor in the extinction of megafauna at the Pleistocene/Holocene transition in Beringia. They show in three ways (a model, an experiment of reintroduction of Yakut Horses and a literature review) that the presence of large herbivores promotes the presence of steppe ecosystems (with many grasses and drier soils), and that their absence promotes a tundra ecosystem (dominated by mosses and with wetter soils). More than the simple climatic conditions, it is these changes in the presence of herbivores that explain the passage from the rather dry grassy steppe to the humid mossy tundra at the end of the Pleistocene. Unlike Soulé and Noss (1998), Zimov *et al.* (1995) clearly aim to recover an ecosystem: the steppe. The reintroduction of large herbivores (and not large carnivores as in North America) is a means to achieve this. They take as an example the successful (re)introductions of Muskoxen in northern Siberia and northern Alaska (Klein (1988) cited in Zimov *et al.* (1995)), of bison in the central mountains of Alaska (Guthrie (1990) cited in Zimov *et al.* (1995)), and recall that Mammoths survived up to 3,700 years BP in the north of Beringia (even though they did not name it Wrangle Island – Vartanyan *et al.* (1993)). It is necessary to restore these steppe ecosystems which have been impacted by humans more than any other ecosystems in the world by agriculture and over-hunting of native species. For them, the Russian far north is the ideal candidate for such restoration and there are still remnants of steppe on the southern and dry slopes. Zimov *et al.* (1995) mention the successful introduction of Yakut horses into the Kolyma tundra (100 km south of the Arctic Ocean) and the translocation of 25 young-of-the-year foals (500 km south of the ocean), which have acclimatized well and are reproducing. The "Pleisocene Park" was created in 1977, but the rewilding operations through the (re)introduction of species started in 1988. Today (according to the park's official website: https://pleistocenepark.ru/), the park hosts 15 Kalmykian cows, 5–15 elks, 3–4 muskoxen, a European bison, 12 American bison, 20–30 reindeer, 18 sheep, 8 yaks and 40 Yakut horses. These animals are kept in an enclosed area of about 20 km^2.

In Europe, Vera (2018) developed a rewilding project in the Oostvaardersplassen polder (OVP) in the Netherlands as early as the 1980s. Vera's approach to rewilding is very different from the North American approach of Soulé and Noss (1998) and the Russian approach of Zimov *et al.* (1995) in several ways. First, if we trace the history of Vera's approach, we realize that it is very empirical and starts from the need to conserve a particular species. The young polder of OVP (created in 1968) was reopened to the sea and thousands of Greylag Geese came to find refuge there during the breeding season. They soon settled in the marshiest parts of the old polder and, by grazing on the reeds, prevented the development of woody plants. But they also needed the unflooded grassy areas to feed. So Vera's initial question was a very practical one: how do you create grassland areas from scratch to provide favorable habitat for geese? Vera's solution was to introduce large herbivores considered hardy or species equivalent to the old European herbivores that are now extinct. In 1983, 32 Heck's cows (assumed to be Aurochs equivalents), 18 Konik horses and 44 Red Deer were introduced to OVP. These large herbivore introductions are the centerpiece of Vera's rewilding. The success Vera expected was rapid: the large herbivores prevented any development of woody plants and created grassland environments from scratch, ideal for feeding Greylag Geese. Vera found proof that it is indeed the large herbivores that control the vegetation and that they are not content to follow the different facies/stages of succession of a vegetation mosaic. The second point of difference comes from the absence of predators. There are no apex predators at OVP and the entire site is fenced off. The regulation of the introduced herbivore populations is mainly done by the scarcity of food resources, combined with winter rigors, sometimes leading to massive deaths of the OVP herbivores, which do not benefit from any artificial fodder supply. The third major difference with the North American vision (and even the Russian vision if we consider that Zimov's "Pleistocene park" is only an experimental site) is the question of spatial scales. OVP is a very small site (less than 6,000 ha), isolated from other nature nodes because it is inserted into very intensive agricultural landscapes (even though a corridor with another polder is envisaged). The three "Cs" of Soulé and Noss (1998) (cores, corridors, carnivores) are therefore absent from this rewilding: the site is very small, corridors are virtually absent and apex predators such as the Grey Wolf are absent. OVP is more like an experimental site for the effect of grazing on plant dynamics. The only wild dimension of this area is perhaps its self-regulation by the control of vegetation by large herbivores, themselves regulated by the vegetation as a trophic resource (one of the rare interventions is the shooting of some individuals following protests by animal rights activists). According to Vera (see Marris (2009)), and as recalled in the first part of this chapter, past environments in Europe were mosaics of closed and open habitats maintained by herbivores. A quick look at Google Earth, however, reveals that none of this has happened at OVP, which remains a very open 6,000 ha area and not at all a mosaic of closed and open environments. In various interviews (see Marris (2009)), Vera does not exclude the spontaneous return of the Grey Wolf to OVP, but does not envisage its reintroduction.

Beyond these three examples, many concrete rewilding projects have emerged in recent years across Europe, which can be seen as offshoots of Vera's work.

Rewilding Europe is an NGO created in 2011 that proposes to (re)introduce large herbivores into well-defined and often enclosed spaces. As Barraud (2020) rightly points out, these projects are essentially based on the structuring role of the large (re)introduced herbivores and, in this sense, are opposed to the agro-pastoral management classically used elsewhere. We may refer to Barraud (2020), who proposes, in a very detailed and argued manner, examples of rewilding in Europe, notably in Portugal (Côa Valley).

To conclude this part, it can be said that rewilding seeks to reintroduce large animal species, but that four major questions remain: what are the objectives of rewilding? What is the relevant time scale? What species should be reintroduced? How can it be done concretely? When reading the scientific production on rewilding, it is rather curious that the final objectives of rewilding are not always clearly explained by their authors. It may be a question of recovering: (1) environments: like the steppe in Siberia, a forest that is either closed or mosaic maintained by large herbivores in Europe; (2) an autonomous functioning, betting on the development of certain environments more than by issuing hypotheses based on ecological theory (for example, OVP in the Netherlands); (3) a somewhat mythologized and idealized nature with a certain integrity (recovering a procession of extinct species considered indigenous and representative of the site), stability (promoting functioning without human intervention) and beauty (nature must regain its splendor). This last point corresponds rather to the North American model, but we also read through the lines of Zimov *et al.* that it is necessary to restore the great steppe (in particular to fight against global warming) and nothing else than the great Russia can allow that. The question of the baseline period is not clear-cut either. However, in North America as well as in Russia, there seems to be an agreement around the end of the Pleistocene and the beginning of the Holocene. This is perhaps less clear for Europe. The candidate species for (re)introduction are also different according to the authors: large carnivores in North America, large herbivores in Europe and Russia. Extinct species can be replaced by contemporary wild functional equivalents (African species to be introduced in America), as well as domesticated species resembling the extinct species (horses, cows, sheep), or even wild species still present (bison, elk, etc.). The way of proceeding is still rather experimental and mainly proposes (re)introductions in closed areas in North America, as well as in Europe and Russia. The objective of these "parks" is to acquire better ecological knowledge for larger projects and free themselves from the constraints of cohabitation with humans and their contemporary activities. In general, it is therefore very difficult to declare what should ideally be done in terms of rewilding, so most of the articles, as we will see in the last part of this chapter, are criticisms, often negative, of rewilding.

5.5. Criticisms and controversies around rewilding

The literature on trophic rewilding is overwhelmingly dominated by opinionated essays and articles at the expense of empirical studies (experimental and non-experimental) and literature review articles (Svenning *et al.* 2016). As a result, the criticisms and controversies raised by rewilding are extensive.

5.5.1. *Ecological criticism*

The main criticisms of trophic rewilding come from the lack of ecological knowledge on which it relies. Svenning *et al.* (2016) point out the lack of empirical studies on rewilding and the fact that knowledge about it is still very fragmented and geographically biased. William Sutherland, referring to rewilding experiments, even quips: "*it is as if you can only find out about details of a cancer treatment by going round the wards with the doctor*", as he believes that those who practice rewilding are too busy with their actual conservation tasks to track and publish their results (Marris 2009).

The first question is whether the species to be established were already present in the past: in other words, is it a question of reintroducing or reinforcing populations, or of introducing them? As we saw at the beginning of this chapter, much of the information concerning the spatial and temporal distribution of extinct species is based on archaeological remains that are sometimes scarce. In the case of an introduction of a species that was never present, can we then speak of "rewilding"?

From a functional point of view, there are also many questions. Will the (re)introduced species compete with species already present? For Rubenstein *et al.* (2006), re-introducing Pleistocene fauna could have catastrophic effects on native species, including species of high conservation concern (see Smith (2005)). For example, the introduction of camels into Australia was catastrophic because they fed on rare native plants (see Edwards *et al.* (2001)). Beyond competition, (re)introduced species can also carry pathogens that are dangerous to native species (Viggers *et al.* (1993) cited in Rubenstein *et al.* (2006); Cunningham (1996) cited in Rubenstein *et al.* (2006)).

The principle of ecosystem functioning controlled by the top of the trophic chain (top-down) is also questionable (see Nogues-Bravo *et al.* (2016)), because, for other authors, ecosystems can be, on the contrary, controlled by the bottom of the trophic chain (bottom-up). Similarly, the engineer species (Jones *et al.* 1994) are species that structure ecosystems, but not by their trophic role. From the point of view of the functioning of all ecosystems, according to Rubenstein *et al.* (2006), we already know so little about the functioning of Pleistocene ecosystems that it would be very difficult to know whether the (re)introduction of extinct or extinct species would restore these extinct ecosystems or create new disturbed ecosystems.

For some authors, the success of (re)introduction of long extinct species seems utopian and illusory, as the reintroduction of extinct species within their own range is already difficult (see the difficulties of reintroduction of Przewalski's Horse *Equus ferus przewalskii* (Wakefield *et al.* (2002) cited in Rubenstein *et al.* (2006)), the Onager *Equus hemionus* in Israel (Saltz & Rubenstein (1995) cited in Rubenstein *et al.* (2006)) and Grevy's Zebra *Equus grevyi* in the horn of Africa (Moehlman (2002) cited in Rubenstein *et al.* (2006))).

5.5.2. *Ethical criticism*

Much of the literature on rewilding proposes the (re)introduction of "functional equivalents" of extinct or lost species. This raises two ethical problems. The first is that nature could be interchangeable and that species could be reduced to the functions they perform. This "interchangeability" of nature is even in some ways evolutionary nonsense. For example, although molecular data suggest that present-day horses are close to their ancestors (see Weinstock *et al.* (2005)), cheetahs proposed to be (re)introduced into North America by proponents of Pleistocene rewilding are quite different from their Pleistocene ancestors (see Barnett *et al.* (2005) cited in Rubenstein *et al.* (2006)). The second problem is to make people believe that there is no irreversibility in disappearances and extinctions and that techno-science would always allow us to turn things around. This form of prometheism has already been questioned by Callicott (2002) concerning ecological restoration and is even more violently criticized by Rubenstein *et al.* (2006), who compare some rewilding projects to "Jurassic parks".

Second, since budgets and study time are not unlimited, how can we legitimize the time and money spent on the (re)introduction of extinct species or the (re)introduction of functional equivalents rather than those whose populations are still viable?

Finally, a barrage of criticism from animal welfare advocates has been directed at some rewilding projects. The Dutch OVP project, which, it should be recalled, introduced hordes of horses onto this polder, relies on regulating these populations through mortality due to harsh climate and food shortages. The managers did not provide the animals with fodder in winter, which led to high animal mortality. Animal welfare advocates then brought hay to the animals over the polder's fence. In 2010, a memorial dedicated to the animals that died at OVP was even erected. More generally, this intervention by animal rights activists raises the paradox of maintaining a population of animals that the manager wants to be wild when they are farmed and live in an enclosed space. Can we qualify something generated from scratch by humans as wild?

5.5.3. *Social criticism*

The (re)introduction of long-extinct large predators or their functional equivalents in a rewilding logic also poses social problems. We already know that the spontaneous return of large carnivores in Europe (see Chapron *et al.* (2014)) pose many problems of conflict with many professions and that their reintroduction, even in areas from which they never disappeared, is conflictual. How then can the introduction of non-native species be accepted as functional equivalents, such as cheetahs or lions from Africa in the United States? For Rubenstein *et al.* (2006), such introduction projects would generate hostility towards conservation measures.

5.5.4. *Cultural and political criticism*

As Barraud *et al.* (2019) well-documented, rewilding also faces significant criticism from advocates of cultural landscapes (as they may be defined by Sauer (1925)). For Barraud *et al.* (2019), the approach of rewilding advocates can be seen as too "speculative" (especially the passive form of rewilding through farmland abandonment). Citing Poux *et al.* (2009) and also Rotherham (2014, 2015), Barraud *et al.* (2019) recall that according to these authors, rewilding could jeopardize agro-pastoral systems already weakened by climate change and the decline of traditional agro-pastoral practices.

Finally, Barraud *et al.* (2019), drawing on the work of Jørgensen (2015), remind us that rewilding raises questions of power:

> Which actors formulate the rewilding project? Which of them participate in the decision? Do these projects correspond to an inclusive form of conservation, responding to the expectations of the inhabitants, local users and local communities? Do they contribute to offer an alternative to the extractivist threats? Who are – in fine – the winners and losers during the deployment of the MNSE [Movement for the Promotion of Wild Nature in Europe] discourses and then during the implementation of concrete rewilding operations?

In conclusion of this part, it can be said that rewilding has provoked numerous criticisms on ecological, ethical, social, cultural and political grounds. Rewilding is not, however, a very recent idea in conservation biology, since it emerged in the 1980s, *i.e.*, at the very beginning of the discipline. The main criticism of rewilding is that it is not based on a strong theoretical consensus and that it does not benefit from sufficient empirical results. The proposed experiments are therefore judged to be fanciful, risky and even dangerous. The criticisms made in the literature have often taken a rather vehement and personal turn towards the proponents of rewilding. Moreover, some proponents of rewilding, such as Vera, have been violently insulted in recent years, especially by animal rights activists.

5.6. Conclusion

Humans have gradually "defaunated" the entire planet Earth. Various studies allow us to trace in broad strokes what Callicott (2002) refers to as an "ecological holocaust". A very large part of the carnivorous and herbivorous megafauna was eradicated at the end of the Pleistocene, at least in part by humans. Mammals and birds have been decimated all over the globe throughout the Holocene and especially during the last centuries. This pressure is even stronger in island environments where the introduction by humans of other species and the overexploitation of those present have led to the extinction of many endemic birds and mammals. In addition, it has been known for some years now that many other taxonomic groups (notably insects and amphibians) are threatened on a global scale and that even common species (widely distributed, abundant and occupying a wide range of habitats) are experiencing massive declines in abundance. The contemporary landscapes that surround us are thus akin to a stage set without its main actors, the animals. Certainly, these ecosystems function in one way or another and we are even able to get used to them (see Chapter 4). Even now, many species are associated with environments that are merely surrogate habitats for them (see Godet (2017)) and a few very generalist species will always pull through (see Olden and Rooney (2006) for more on the concept of biotic homogenization) so that an illusion of living landscapes remains. But isn't accepting this resignation tantamount to supporting an inexorable decline in biodiversity? Timid proposals for a return to the past are proposed in the conservation world, such as the return to the agro-pastoral landscapes of yesteryear, often a little idealized and above all quite consensual in the sense that they correspond to the logic of sustainable development linking the economic, social and environmental. But faced with the urgency of the crisis and the desolate state of the biosphere, rewilding does have a broader ambition. Not dealt with in this chapter, but in Chapter 6 of this book, rewilding can take place through spontaneous dynamics. Some species, such as large predators, are naturally returning to Europe (Chapron *et al.* 2014) and, at the time of writing, the arrival of the Golden Jackal in the Bouches-du-Rhône (southern France) is being reported. Similarly, some areas benefit from agricultural abandonment and become enriched (see Génot and Schnitzler (2012) for France). Active rewilding, on the contrary, makes it possible to re-invigorate the dynamics and functioning of ecosystems as they were before the omnipresent action of humans on Earth by (re)introducing large herbivores and/or large carnivores. However, this comes up against multiple misunderstandings that we will not list again in this conclusion. In general, it can be said that we have become so accustomed to living in narrow, lifeless landscapes that we find it very difficult to rethink them with any historical depth, for example, as being traversable and controllable by large species other than ourselves. The very world of conservation scientists has even turned so much to an anthropocentric and utilitarian view of nature conservation that restoring self-regulated living spaces seems inconceivable. Experiments in trophic rewilding have thus far been confined, at best, to small, fenced-off sites, or, often, to mere ideas on paper. Of course, we do not know everything about the future impact of (re)introductions of extinct species, but should we wait until we know everything before acting? Conservation, in essence, like other disciplines such as medicine, often proceeds empirically. However, we lack so much data to validate or invalidate the interest of rewilding that it is impossible today to reject this proposal on serious grounds. In our view, "species de-extinction" projects, such as those for the Woolly Mammoth (see McCauley *et al.* (2016)) or the translocation of species deemed to be "functional equivalents" of extinct species (*e.g.*, introducing elephants to the United States) are undoubtedly a matter of our species' *hubris*. The potential effects of such a strategy would be disastrous in several ways, but above all because it would make us believe in the omnipotence of technology to repair our mistakes. However, we cannot solve a crisis by using the methods that led to it. On the contrary, repairing our recent errors of the last few centuries seems reasonable and feasible. An acceptance of the return of the wild and the reintroduction of disappeared species in non-enclosed spaces of an indigenous fauna, seems completely conceivable and increases the palette of possibilities. Indeed, the small enclosed spaces that have been rewilded with species that are often domesticated, although they can be interesting experimental grounds for understanding some ecological functions, are not the key to a true conservation policy. How can one speak of nature, for example in Oostvaardersplassen, on a fenced polder of 6,000 ha where domestic animals are introduced and which is devoid of any predators? On the contrary, rewilding projects that combine the reintroduction of large carnivores and large herbivores that only disappeared during the last centuries in large areas of agricultural and rural abandonment are a reasonable solution. Protected areas, as the keystone of nature conservation, have shown and still show their effectiveness. However, they should only be temporary and emergency measures. The areas of abandonment in which large carnivores and herbivores have been reintroduced are artificial only in their creation and in the introductions during their "launching". They do not exclude humans. They simply constitute spaces beyond the space inhabited and controlled by humans. As we write these lines, we can already imagine the critics of a romanticized and nostalgic vision of the promotion of an idyllic nature that may never have existed. The opposite is true. These spaces from which humans have withdrawn and where they have left control to other large vertebrates are the spaces of tomorrow, whose functioning and new abundance of life that may take place there are not yet known. Let us encourage the creation of such spaces and let them finally function outside of our species' control and regulation. One day, I hope to hear the following discussion from the mouths of our children:

> Where are you going on vacation this summer?
>
> I'm going on safari in France!
>
> Take a map so you don't get lost and don't leave food around your tent at night!
>
> Yes, I know, the Golden Jackals are mischievous!
>
> Not less than the Brown Bears in any case …
>
> And take your binoculars, I hear there are White-backed Woodpeckers in the beech forest.
>
> So they reintroduced them?
>
> No no, they came back by themselves.

6

Spontaneous Rewilding through Land Abandonment

6.1. Introduction

Globally, wilderness areas have suffered a catastrophic decline in the last century: 3.3 million km^2 of environments not modified by humans disappeared in the two decades, beginning in 1990 (Watson *et al.* 2016). In Europe, where the human footprint is long established, these spaces are now only relict: for forest environments, Sabatini *et al.* (2018) estimate that only 0.87% of European forests can be considered primary. In metropolitan France, only 29% of forests are ancient and less than 3% of managed woodlands significantly exceed the age at which they would normally be harvested (Cateau *et al.* 2015).

However, in parallel with the progressive disappearance of these wilderness areas in France and Europe, other natural spaces are appearing spontaneously. These are areas where anthropogenic pressures have weakened, *i.e.*, spaces where the human hold has more or less recently loosened and which are sometimes described as novel ecosystems (Hobbs *et al.* 2013), of a feral nature (Génot and Schnitzler 2013), spaces of naturalness (Guetté *et al.* 2018), and which constitute a form of third landscape (Clément 2004). All of these spaces, on the contrary, share the common characteristic of having self-sustaining ecological processes intact.

The appearance of these "free" spaces may be the result of what is known as "rewilding". While rewilding is more often understood as the active process aiming to restore an ecosystem to a character and functionality similar to what it could have had before anthropic domination (see Chapter 5), it also includes another form, which is a spontaneous process that often appears in spaces previously used by

---

Chapter written by Adrien GUETTÉ and Jonathan CARRUTHERS-JONES.

humans. Tsing (2017) distinguishes rewilding, which refers to the introduction of animals by humans into disturbed landscapes, from auto-rewilding, which corresponds to rewilding by animals, plants and other organisms themselves (re-wilders). This second form of rewilding, which can occur in a variety of contexts (Pic *et al.* 2020), more or less voluntarily and more or less suddenly, is sometimes referred to as "passive" (Pereira and Navarro 2015). We will refer to it in this chapter as "spontaneous rewilding".

One of the unique aspects of spontaneous rewilding is the absence of reference to a past state. In contrast to certain conservation measures that often aim to maintain or recover a known ecosystem or landscape "state" (such as trophic rewilding), spontaneous rewilding is not anchored in the past and does not seek to return to it. The only reference is that of the spontaneity of processes that may have existed in the past in the same space, but whose form will be unique. In Europe and France, the largest areas of spontaneous rewilding are probably found in areas of land abandonment (Carver 2019). And since the trend of abandonment of agricultural activities in many European regions seems to be increasing (Navarro and Pereira 2012; European Commission 2018), these spaces will quickly find themselves at the heart of debates on socio-economic and ecological issues. In France, abandoned agricultural lands wastelands are relatively poorly documented and their national mapping is non-existent, so that reflections on their interest and future are mainly decided on a sectoral and local scale. We believe that a better knowledge of their number, surface and distribution would help to better consider the importance and value, particularly ecological, of these spaces, which have the potential to contribute to the mitigation of some of the negative effects of global landscape change, provided they are left to develop freely.

In the first part of this chapter, we will see how land abandonment can be considered as a form of spontaneous rewilding. In the second part, we will propose a study of its dynamics at the scale of France between 1990 and 2018. In the last part, we will discuss the issues of identification and recognition of these rewilding areas.

6.2. Land abandonment: a form of spontaneous rewilding

Land abandonment is the accepted term for the process of replacing agricultural use of land with that of another. It is a phenomenon that has affected all of Europe since the 19th century (Lasanta *et al.* 2017). The reasons for the cessation of agricultural use are mainly related to the erosion of soil fertility, changes in the economic structure of farms and regional demographic factors (Keenleyside *et al.* 2010; European Commission 2018). While this phenomenon is often viewed and experienced negatively, especially by local populations, as it reflects a certain desertification of the countryside, it also represents an opportunity for spontaneous rewilding (Navarro and Pereira 2012).

6.2.1. *Active versus spontaneous rewilding*

According to Aplet *et al.* (2000), spaces can be characterized according to two basic qualities. A first axis of "naturalness" describes the ecosystem's integrity, while a second axis of "wildness third landscape" describes the degree to which the space remains outside of human control. These two qualities are distinguished by reference to a known state. While ecosystem naturalness is measured against a historical standard, the degree of wildness is only relative to contemporary human control over the ecosystem. Thus, some areas may be natural but not free, as in the case of certain protected habitats that require the maintenance of agropastoral activity; others may, on the contrary, be considered as not very natural but free. It is in this last category of spaces with low ecological naturalness but high wildness that we can classify different types of spaces that appear following spontaneous rewilding, such as "novel ecosystems", feral spaces, wastelands or the third landscape (see Figure 6.1), which we will present in the following section.

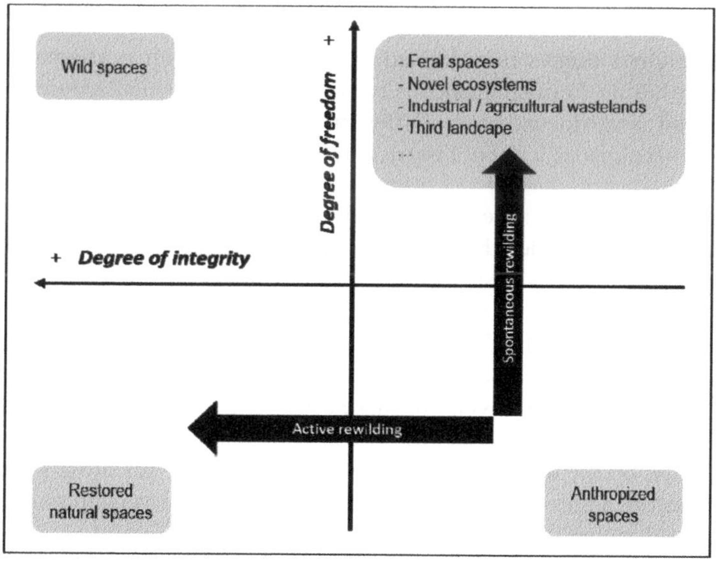

Figure 6.1. *Conceptual diagram of "active" and "spontaneous" rewilding processes in terms of the qualities of ecosystem naturalness and process wildness*

These two qualities of "naturalness" and "wildness" are, in fact, at the heart of a debate still active among conservation biologists. According to Ridder (2007), the debate between naturalness and wildness is mainly about values: on the one hand, the desire to conserve biodiversity, and on the other hand, the desire to minimize direct human intervention on nature. This is probably what most distinguishes the two forms of active and spontaneous rewilding. The aim of the former is to achieve an ecosystem and associated biodiversity whose baselines are known and defined, whereas the latter aims above all for the autonomy of processes, without reference to a state or structure. According to Schnitzler *et al.* (2008a, 2008b), it is the acceptance of new trajectories and new specific assemblages that result only from natural processes, regardless of the history of human activities that preceded and influenced them. The baseline in this case can be the comparison with other sites considered very spontaneous. Thus, as summarized by Landres *et al.* (2000), the dilemma of ecological restoration or active rewilding is that it mechanically implies a decrease in the wildness of processes due to the human action of rewilding. There is thus an opposition and even a contradiction between active rewilding and spontaneous rewilding.

6.2.2. *New open spaces resulting from spontaneous rewilding*

Unlike restored ecosystems defined by comparison with a known past state, ecosystems resulting from spontaneous rewilding have the singularity of being less well described and documented. Different notions and neologisms have been dedicated to these spaces without historical references for some years.

These spaces can first be assimilated to "novel ecosystems". The concept emerged in the 2000s under the influence of Richard Hobbs. Hobbs *et al.* (2013) define novel ecosystems as systems composed of biotic, abiotic and social elements (including their interactions) that, due to human influence, differ from historically prevalent ecosystems and tend to be self-organizing and exhibit new qualities. For its critics, the concept is poorly defined and can lead to undesirable practices because by recognizing the novelty of these ecosystems, on the one hand, it ignores all the work that has been done for decades on anthropogenically transformed ecosystems and, on the other hand, it conveys a message of scientific approval that the new (therefore anthropized) ecosystems are an improved version (since they are new) of an "old nature" and associated study methods (Murcia *et al.* 2014). These criticisms come primarily from the restoration ecology community, some of whose members fear that the novel ecosystem concept will supplant ecological restoration initiatives in conservation policy, particularly for economic reasons. Proponents of ecological restoration believe that "labeling" novel ecosystems involves no intellectual effort and is not based on any empirical evidence, whereas ecological restoration success stories abound. In between these two views, Marris *et al.* (2013)

propose that the concept of a "novel ecosystem" serves primarily as a transitional tool to encourage education around the areas of human disturbance, sometimes pejoratively referred to as "trash ecosystems". According to Marris *et al.* (2013), naming these novel ecosystems is simply an acknowledgement of their existence and present interest. It is not a matter of calling all ecosystems "novel" under the pretext that the entire planet is under the direct or indirect influence of humans, but rather of assigning value to the self-determined natural process and those ecosystems that do not resemble or no longer resemble their distant ancestors. According to the authors, this notion may present an opportunity for restoration ecologists to throw off the yoke of history, because as we restore ecosystems that are increasingly rare and removed from real-world conditions, a shift occurs towards ecosystem "design" (Marris *et al.* 2013).

Other terms and notions have been used to describe these ecosystems or habitats born from spontaneous wilderness. Using the term "feral" to designate domesticated species returned to the wild, some authors have extended its application to landscapes or territories, which after a human history have been abandoned, thus regaining a wild character. According to Schnitzler and Génot (2012), since a good part of Europe, and almost all French landscapes, have a multi-millennial human history, it is in the feral spaces, *i.e.*, the spaces where the human hold has been loosened, that a rebirth of the wild is possible. A wilderness that is not that of pre-human ecosystems, but whose autonomous character is stronger than that of restored and managed ecosystems. According to these same authors (Génot and Schnitzler 2013), the feral wilderness is an opportunity to rewild France and a sign of confidence in the future. Vast areas of wilderness already exist; we just need to let nature do its thing, as it knows better than humans what is best. The best strategy for rewilding is to give it space and time.

These spaces abandoned by human influence are also sometimes referred to by other expressions. They correspond, for example, to the usual definition of wastelands, which designate spaces abandoned from their primary vocation, primarily industrial. These spaces can also be associated with what the French landscape architect Gilles Clément calls "*Tiers-paysage*" or third landscapes, which he defines as "abandoned land" that follows the abandonment of land formerly exploited by humans and that has the characteristic of having a strong spontaneous dynamic (Clément 2004). According to Mathevet (2004), these are also spaces of "spontaneous nature" (whose functioning owes nothing to humans) and of a "recomposed nature" (that which has developed spontaneously within an anthropized space). This also corresponds to "ordinary" nature as defined by Godet (2010). This quality of "spontaneous processes" also corresponds to one of the three facets of landscape naturalness (in addition to biophysical wildness and spatio-temporal continuities), as defined by Guetté *et al.* (2018).

6.2.3. *Encouraged or tolerated spontaneity*

These spaces, which have a degree of process spontaneity in common, can, however, be separated into two categories linked to their context of appearance. On the one hand, we can distinguish between spaces that have been voluntarily withdrawn from human control and management; this is a spontaneous process that can be described as "encouraged" or "intentional" in the sense of Marty (2005), and on the other hand, spaces whose spontaneity has been imposed progressively or suddenly in territories formerly controlled by human activities. This rewilding can be described as "tolerated" since it is not initiated by the inhabitants of the territories concerned.

"Encouraged rewilding", which leaves room for the natural evolution of landscape dynamics, is still marginal in Europe. The issue of non-intervention as a management measure is gaining ground, but is currently limited to a few protected areas (Rouveyrol *et al.* 2019). Even in these protected areas, the implementation of rewilding strategies poses numerous ecological and psychological reticences among land managers (Génot 2008). Yet, a number of socio-ecological benefits can be attributed to rewilding and rewilded landscapes. Landres (2010) mentions six of them in particular: the opportunity to deepen respect for the autonomy of nature, develop scientific humility in the face of a relatively poor understanding of ecological systems, better accept evolutionary change, limit species-centered management and reduce its unintended harmful consequences, create new baselines of uncontrolled habitats and preserve future options. In France, the few rewilding initiatives are mainly private initiatives. Through the *Programme Régional d'Espaces en Libre Evolution* (PRELE), the *Conservatoire d'Espaces Naturels Normandie* has set itself the mission of developing a network of rewilding sites. The *Association pour la Protection des Animaux Sauvages* (ASPAS) has also taken up the objective of acquiring land through participatory financing campaigns to prohibit all human activities and promote natural processes over human management of ecosystems. The label *Réserve de Vie Sauvage*® is granted to "natural spaces whose management strategy is non-intervention, spontaneous or passive". We can regret the private character of these initiatives, but in the absence of a public policy on rewilding, they offer a showcase of what free nature is in the French geographical context.

"Endured rewilding" is distinguished by an involuntary withdrawal of human presence and/or activities. Rewilding in this case is only the subsequent effect. This form of rewilding can occur as a result of a sudden event such as a disaster or, on the contrary, as part of a slow and progressive dynamic. One of the most striking examples of post-disaster rewilding is undoubtedly the one concerning the deserted territories of Belarus and Ukraine following the explosion of a reactor at the Chernobyl nuclear power plant on April 26, 1986. For more than 30 years now, nature has evolved freely within the 4,200 km^2 of the exclusion zone where the 116,000 inhabitants were evacuated to protect them from radionuclide contamination. This human tragedy was also a unique opportunity to measure the effect of spontaneous rewilding on biodiversity. A study conducted on mammal abundance suggests that the negative effect of radiation is largely offset by the positive effect of removing persistent and ever-increasing anthropogenic stress on ecosystems (Deryabina *et al.* 2015). The Covid-19 pandemic is another example of a human disaster that will likely impact the spontaneous rewilding of some territories. Scientists are already mobilizing to study the effect of this "anthropause" (Rutz *et al.* 2020) on wildlife. It is likely that the abandonment of certain areas, whether to reduce the risk of transmission of zoonoses (diseases that pass from wild animals to humans) or as a result of the economic crisis that the pandemic could bring in its wake, will take place over long periods of time.

Less tragic than the previous examples, land abandonment is also part of this dynamic of spontaneous and suffered rewilding. The processes at work are, however, slower and more gradual, so that their detection requires studies on sufficiently large scales of time and space, while promoting a large geographical scale to measure the dynamics.

The following section proposes to study the phenomenon of spontaneous rewilding linked to land abandonment in France between 1990 and 2018. What were the combinations of land use changes at work, which areas were concerned and where were they located?

6.3. Quantifying and mapping spontaneous rewilding areas related to land abandonment

6.3.1. *Context*

The cheapest and most effective way to promote landscape rewilding is to eliminate or reduce the causes that have contributed to their degradation as much as possible. Yet, over the past 12,000 years in Europe, these causes have been primarily related to agriculture and grazing that have replaced much of the natural vegetation (Rey Benayas 2019).

But since the 19th century, with an acceleration from the mid-20th century, a dynamic of land abandonment affected the whole of Europe (Lasanta *et al.* 2017). During the period 2015–2030, it is estimated that about 11% (or more than 20 million hectares) of European agricultural land is at high risk of abandonment. At any given risk level, as much as 183 million hectares are considered to be at "potential risk" of land abandonment by 2030 (Castillo 2021). The most affected regions appear to be southern and eastern Romania, southwestern France, southern and central Spain, Portugal, Cyprus, Poland, Latvia and Estonia (European Commission 2018).

Various studies have shown the positive and negative effects of land abandonment. Among the negative effects, we identify the loss of agricultural biodiversity and cultural heritage. Other positive effects counterbalance these negative effects, such as the increase in carbon sequestration capacity, the development of nature-based recreation or the extension of available habitats for mammals (Estel *et al.* 2015; Van Der Zanden *et al.* 2017). If we add the various arguments in favor of spontaneous rewilding spaces mentioned in the first part to these positive effects, we understand that the dynamics of land abandonment could be at the center of issues that go far beyond agricultural production and that it is perhaps urgent to collectively question the future of these spaces, which seem to be spreading over increasingly large areas.

Many studies attempt to measure the extent of land abandonment. A first category of studies proposes to quantify the areas that could be affected in the more or less long term by using socio-economic models of land cover and use evolution (Price *et al.* 2015; Van Der Zanden *et al.* 2017; Perpiña Castillo *et al.* 2020; Castillo 2021). A second category of studies based on empirical data is interested in measuring and mapping the areas currently affected by land abandonment, at different scales and concerning different environments: at the European scale (Keenleyside *et al.* 2010; Estel *et al.* 2015; Lasanta *et al.* 2017), regarding mountain spaces (MacDonald *et al.* 2000) or at regional scales, in Spain (Corbelle-Rico *et al.* 2012) and in Italy (Falcucci *et al.* 2007). To our knowledge, the studies concerning France are essentially: either those carried out on a European scale; or based on declarative data (notably the agricultural census and the "land use" survey (TERUTI)) (Pointereau *et al.* 2008); or more local (*e.g.*, Houzard and Lecointe 1991; Arlaud 1994; Bioret *et al.* 1994; Perigord 1994; André 1995; Preux *et al.* 2019). The study of land abandonment at the scale of metropolitan France by using spatially explicit data seems to us to be little explored. In this section, we propose to use land use data to produce a spatially explicit map of areas of land abandonment. The objective is to cover all of metropolitan France, with a geographical resolution of 100 meters, using Corine Land Cover data whose temporal scope extends from 1990 to 2018.

6.3.2. *Material and methods*

To identify the probable areas of land abandonment, we have retained the definition most commonly used by the European Commission. Land abandonment areas are "areas previously used for cultivation, grazing or farming that no longer have agricultural functions and that have not been converted into forests or artificial areas" (Pointereau *et al.* 2008; Castillo 2021).

We used CORINE Land Cover (CLC) data because they have the advantage of covering the whole of metropolitan France across nearly 30 years (1990–2018). Data from the *Registre Parcellaire Graphique* (RPG), which have a better spatial resolution, were excluded from our analysis because the available time series is shorter (2010–2019) and the data only represent the areas used as a baseline for the Common Agricultural Policy (*Politique Agricole Commune*, PAC) subsidies (Cantelaube and Carles 2014). The CLC nomenclature includes 44 land use classes. To reduce the risk of error related to photo interpretation, which was measured at 18% for the year 2000 (Büttner *et al.* 2006), we used level 2 of the nomenclature, which distinguishes only 15 land use classes (see Table 6.1).

Level 2	CLC class	Code*
1.1	Urban fabric	ZU
1.2	Industrial, commercial and transport units	ZI
1.3	Mine, dump and construction sites	MINE
1.4	Artificial, non-agricultural vegetated areas	EV
2.1	Arable land	TA
2.2	Permanent crops	CP
2.3	Pastures	PRAI
2.4	Heterogeneous agricultural areas	ZA
3.1	Forests	FORET
3.2	Scrub and/or herbaceous vegetation associations	VEG_A
3.3	Open spaces with little or no vegetation	EO
4.1	Inland wetlands	ZHI
4.2	Marine wetlands	ZHC
5.1	Inland waters	EC
5.2	Marine waters	EM

*The codes provided here have been derived from the French version of the CLC classes.

Table 6.1. *Level 2 CLC nomenclature (source: https://land.copernicus.eu/user-corner/technical-library/corine-land-cover-nomenclature-guidelines/docs/pdf/CLC2018_Nomenclature_illustrated_guide_20190510.pdf)*

We first analyzed the different combinations of land use change between 1990 and 2018. The areas were calculated and represented by a Sankey diagram to illustrate the land use transfers between the two dates. To measure and map likely areas of land abandonment, we then selected combinations of land use change, characteristic of the abandonment of an agricultural activity. The combinations selected were as follows: "arable land" to "scrub and/or herbaceous vegetation associations", "permanent crops" to "scrub and/or herbaceous vegetation associations", "pastures" to "scrub and/or herbaceous vegetation associations" and "heterogeneous agricultural areas" to "scrub and/or herbaceous vegetation associations".

Finally, these combinations were mapped. The gross surface areas per department were calculated and the density of areas of abandonment per km^2 was calculated by a moving window method, in order to produce a more readable map of the territories of land abandonment.

Land use change analyses were performed with the OpenLand (Exavier and Zeilhofer 2020) R package and mapping with ArcGis Pro.

6.3.3. *Results*

6.3.3.1. *Land use changes between 1990 and 2018*

Between 1990 and 2018, 225 (15^2) combinations of land use change and non-change were calculated. The largest land use changes in decreasing order of area were "heterogeneous agricultural areas" to "arable land" (7,078 km^2), "pastures" to "arable land" (6,593 km^2), "forest" to "scrub and/or herbaceous vegetation associations" (6,117 km^2) and "arable land" to "heterogeneous agricultural areas" (5,955 km^2). The total area covered by the analysis was 532,223 km^2.

6.3.3.2. *Areas of land abandonment*

Between 1990 and 2018, the areas corresponding to the four combinations identified as being probably linked to land abandonment represent, in order of importance, 385 km^2 for the combination "pastures" to "scrub and/or herbaceous vegetation associations", 379 km^2 for "heterogeneous agricultural areas" to "scrub and/or herbaceous vegetation associations", 169 km^2 for "arable land" to "scrub and/or herbaceous vegetation associations" and 106 km^2 for "permanent crops" to "scrub and/or herbaceous vegetation associations". The total area of agricultural wasteland is 1,041.12 km^2.

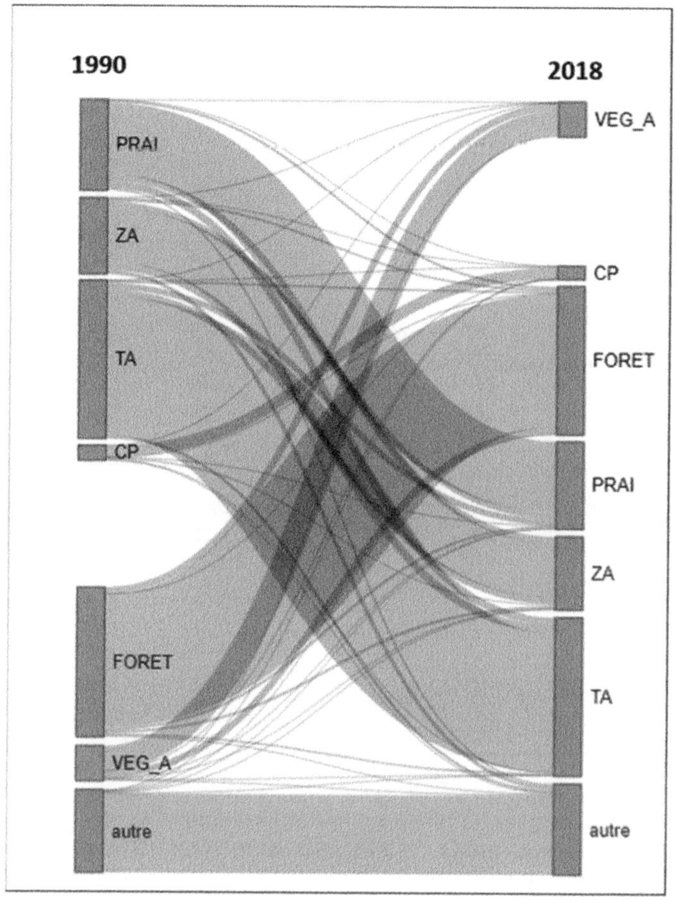

Figure 6.2. *Sankey's diagram representation of land use area transfers between 1990 and 2018. For this representation, land use types other than PRAI, ZA, TA, CP, VEG_A and FORET (see Table 6.1. for acronyms) were grouped into the "other" category. For a color version of this figure, see www.iste.co.uk/godet/baselines.zip*

6.3.3.3. *Spatial distribution of areas of land abandonment*

On the scale of metropolitan France, the areas of land abandonment are not evenly distributed. They are mainly located in the Massif Central, the eastern part of the Pyrenees, the Landes, the Mediterranean arc, Corsica, the Alps, and to a lesser extent, in Brittany. Aude is the department with the largest area of abandoned land (76 km^2) and three departments in Ile-de-France have no abandoned land at all. The departmental average is 11.9 km^2 (standard deviation 15.20 km^2) and the median is 5.06 km^2 (see Figure 6.3).

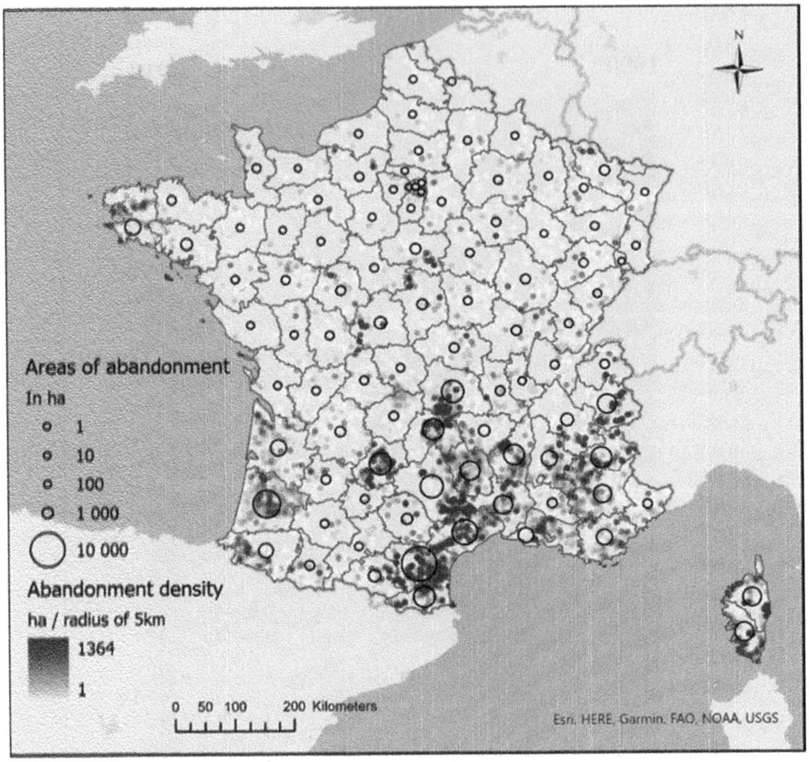

Figure 6.3. *Mapping of areas of land abandonment. The proportional circles represent the gross surfaces in hectares per department. The red gradient indicates the density of abandoned areas (in ha) calculated by a moving window of 5 km radius. For a color version of this figure, see www.iste.co.uk/godet/baselines.zip*

6.3.3.4. *Discussion of the results*

The total area of land abandonment measured appears to be much smaller than other studies conducted in France. Pointereau *et al.* (2008) estimated the area of land abandonment between 1988 and 2000 to be 9,365 km^2. Various explanations can justify this difference. First, Pointereau *et al.*'s study is based on censuses that have the advantage of detecting abandonment before it is visible on the landscape, unlike our approach, which can only take into account land abandonment that is sufficiently old for the landscape dynamics to be detectable via remote-sensing. Indeed, with the time span of our study being 30 years, we can assume that our approach allowed for the detection of abandonments that mainly appeared between 2000 and 2010. Abandonments of less than 10 years are probably undetectable since it is believed that it takes 10–50 years to go from a pioneer stage to a shrub heath. Conversely, the oldest abandonments (between 1990 and 2000) may have already evolved to forest stages, depending on climatic, pedo-topographical conditions, which were excluded from our analysis. This argument is illustrated in Figure 6.4, which traces the landscape dynamics of an area of land abandonment detected by our analysis.

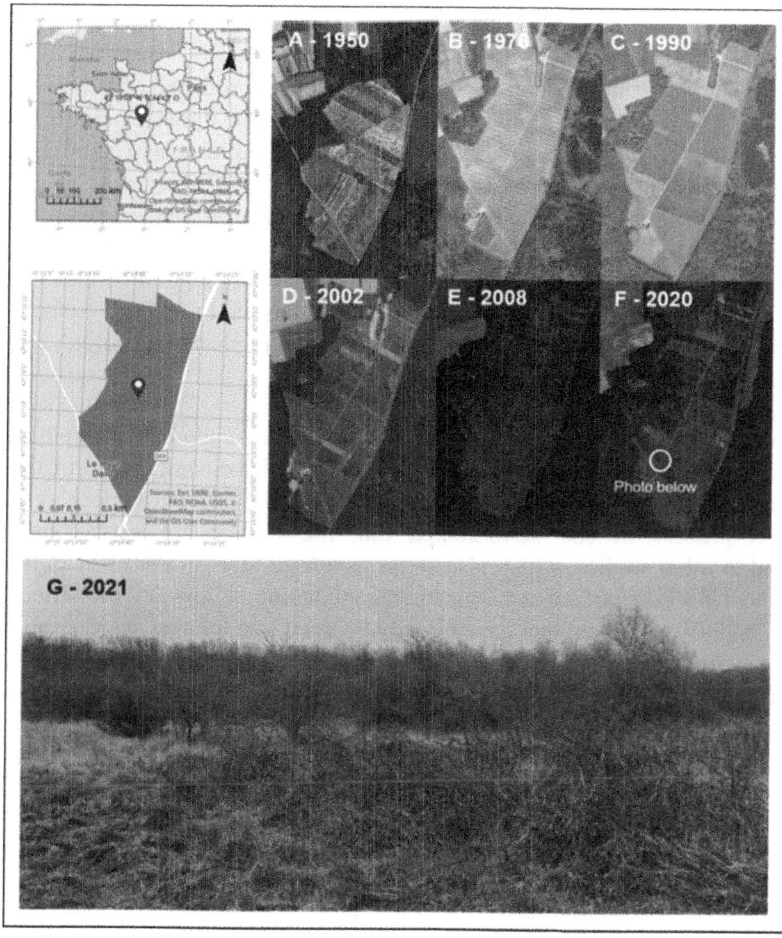

Figure 6.4. *Landscape dynamics of an area of land abandonment detected by cartographic analysis. This is an area of 30 hectares located 5 km south of Gennes (Maine-et-Loire, France). The analysis of aerial photographs reveals that from 1950 to 1990 this area saw different agricultural uses (photos A, B, C). From the 2002 photograph, we observe a fragmentation accompanied by the abandonment of certain sectors (photo D). In 2008, this trend continues with the development of a shrub layer indicating the prolonged absence of agricultural use (photo E). By 2020 and 2021, most of the area appears to be abandoned from agricultural use. The diversity of plant strata indicates a gradual withdrawal of agriculture over the period (photo F and G). For a color version of this figure, see www.iste.co.uk/godet/baselines.zip*

Moreover, the study by Pointereau *et al.* stopped in 2000, whereas our analysis period ran until 2018. We can hypothesize that some of these areas of abandonment were artificialized between 2000 and 2018, knowing that from 2009 to 2014, the annual rate of land artificialization in France was between 9 and 10% (Kraszewski 2019).

From a spatial point of view, we observe a good concordance of our results with other studies, particularly European ones, concerning the Pyrenees and the Massif Central (Estel *et al.* 2015), the Alps and the Mediterranean arc (European Commission 2018). The concordance is even stronger with the results of the study by Carolina Perpina Castillo and her colleagues at the European Commission's Joint Research Centre, on fine-scale modeling of areas of abandonment in Europe. The departments for which the models predict significant depopulation correspond fairly closely to the departments that emerged from our analysis. In addition to the regions mentioned in the other studies, certain departments were clearly identified, including Aude, Landes, Gironde and Finistère (Castillo 2021).

It can be concluded that the proposed methodological approach has a satisfactory capacity to map areas of land abandonment. As for the quantification of the areas concerned, although the work would have to be extended by a more systematic validation process, it can be assumed that the figures are very conservative compared to other studies, which also suggests a high level of confidence. It should nevertheless be emphasized that this method does not allow for the detection of areas of land abandonment that have already evolved into a forest stage, even though these areas probably represent significant surfaces, and the ecological stakes relative to spontaneous afforestation are important. The choice to exclude these areas is based on the possible confusion between new forestry areas and spontaneous forest areas in the data used. In the absence of better data, and with the precautions mentioned, the proposed spatial approach seems sufficiently robust to be used as is. Further work including other CLC millennial figures in date (2000, 2006, 2012) would be useful to understand the landscape trajectories of these areas of land abandonment more clearly.

6.4. Increasing awareness of rewilding areas

Land abandonment is characterized by the end of agricultural activity. These spaces encompass a wide range of contexts that can lead to the appearance of very diverse environments, depending, in particular, on the duration of human non-intervention, the history of the landscape and past uses, the landscape environment and pedo-climato-topographic conditions. But all these spaces possess a strong autonomy of process, so that they fall within the framework of spontaneous rewilding as we defined it in the first part of this chapter.

Alongside active forms of rewilding that are sometimes considered a radical proposition (Barraud *et al.* 2019) and can only involve limited areas, spontaneous rewilding related to land abandonment has a greater capacity for acceptance and the potential to promote natural ecosystem dynamics over larger areas. Because of its current and future extent, land abandonment at least deserves to be discussed in the framework of public nature conservation policies.

These spaces have several undeniable advantages. First, they are available and are only marginally coveted by other uses, since the socio-economic reasons that led to their abandonment from agriculture also apply – at least in part – to industry or urban development. As we have seen, abandonment sites especially appear in areas that are poorly served and difficult to access, and where the population tends to stagnate or decline.

Second, these spaces represent a significant total surface area. In a densely populated country like France, where all spaces are highly coveted, the availability of a surface area equivalent to 10 times that of Paris deserves a future where these spaces are a focus of public debate. But for the choice to be made to conserve them, we will also need a socio-ecological evolution of our relationship with open spaces.

6.4.1. *Identifying areas of land abandonment before they are urbanized*

Although the trend of increasing areas of land abandonment is considered alarming at the European level (European Commission 2018) and France is not exempt from this structural trend, its mapping remains very poorly documented. One of the explanations for this is the intersectoral nature of land abandonment. No longer truly agricultural, not urbanized and not considered as areas at stake for nature conservation, abandoned agricultural lands escape sectoral inventories and censuses. In the agricultural sector, the loss of agricultural land is certainly documented, but without measuring precisely in what proportions these losses are explained by the extension of urbanization or simply by abandonment (Aillery *et al.* 2018). Only the ONCEA (2014) study offers an estimate based on the losses of agricultural land between 2000 and 2006. Over this period, 93% of the disappeared agricultural land was replaced by artificialized land.

However, as early as 2010, within the framework of the Law on the Modernization of Agriculture and Fishing (*Loi de modernisation de l'agriculture et de la pêche*, LMAP), the objective of reducing the consumption of agricultural land was accompanied by the creation of the *Observatoire national de la consommation d'espaces agricoles* (ONCEA) in 2013. With the Law on the Future of Agriculture, Food and Forestry (*Loi d'Avenir pour l'Agriculture, l'Alimentation et la Forêt*, LAAAF) of October 2014, ONCEA's scope of observation was expanded to include natural and forest areas. ONCEA thus became the *Observatoire national des Espaces Naturels, Agricoles et Forestiers* (OENAF) in 2016. However, this national observatory is based on regional organizations that are struggling to be formalized, and whose observation systems are heterogeneous in their form and purpose (CEREMA 2017). Improving the monitoring of these areas of land abandonment will perhaps emanate from the issue of stopping the artificialization of land, which has been more clearly established through the "zero net artificialization" objective of the 2018 Biodiversity Plan. One of the measures of this plan involves the annual publication of an inventory of space consumption, as well as the provision of transparent and comparable data for territories and citizens at all territorial scales (Kraszewski 2019).

But there has been recent interest, specifically in abandoned land. In June 2019, Emmanuelle Wargon, then Secretary of State to the Minister of Ecological and Solidarity Transition, launched a working group on the restoration of abandoned land, particularly for their potential to produce photovoltaic energy following a study by ADEME (2019). This working group then embraced the theme of the identification and restoration of brownfield sites more broadly. Several proposals emerged from the working group, notably the need to legally define the status of abandoned land and develop territorial inventories of their location[1]. Indeed, it was noted that there is a profound lack of knowledge of abandoned lands (in the broad sense of the term, defined as industrial, mining, agricultural, commercial, residential or even railway) in terms of their number, surface area and distribution. A survey of inventories of brownfield sites was conducted in 2018 by the *Junior entreprise de l'école centrale de Lille* (Monnier and Potier 2018) at the request of Lifti (*Laboratoire d'Initiatives Foncières et Territoriales Innovantes*). This study concluded that there is a significant disparity in initiatives across territories. It should be noted, however, that the purpose of these inventories is overwhelmingly related to the identification of "land availability" (Monnier and Potier 2018). Without a concerted policy, there is a significant risk that abandoned spontaneously rewilding areas will be seen solely as land deposits without considering their ecological characteristics and the inherent environmental issues. And this risk of premature disappearance seems to increase as new interest in these spaces develops without the question of their transformation (or preservation) being asked collectively. In the meantime, we are witnessing the somewhat absurd situation where the transformation of abandoned lands is encouraged and justified by the objective of "zero net artificialization", which aims, in particular, to preserve biodiversity. Indeed, one of the tools that supports the "zero net artificialization" objective of the Biodiversity Plan is the implementation of a mapping of abandoned lands. The Cartofriches project, launched in July 2020 by CEREMA, aims to help

1 Report of the *Commission d'enquête, travaux parlementaires*, accessed March 1, 2020. Available at: http://www.senat.fr/rap/r19-700-1/r19-700-112.html#fn312.

identify abandoned lands that are considered to be "breeding grounds[2]" for communities and project developers. One can seriously question the interest of the Biodiversity Plan in terms of biodiversity preservation if the tools implemented aim to urbanize only those spaces where the human footprint has spontaneously eased.

6.4.2. *(Re)giving a place to open spaces*

The role of open spaces, *i.e.*, spaces in which humans do not directly intervene in ecological processes, is increasingly reduced on a global scale (Watson *et al.* 2016). In France, we may mistakenly believe that protected areas guarantee this autonomy of natural processes. However, of the 17 million hectares of terrestrial protected areas in metropolitan France, only 23,621 ha[3] can truly be considered as spaces where ecological processes can develop naturally. These are the integral biological reserves and the integral reserves of national parks, which are explicitly dedicated to the natural evolution of forest environments and associated environments (INPN 2020). All other protected area statutes tolerate or even encourage human maintenance and extraction activities under certain conditions. With regard to the wildness of natural processes, the areas of land abandonment mapped in the second part of this chapter certainly present a greater naturalness than many French protected areas. Although there is a movement in favor of these open spaces, the establishment of a protection policy is not on the agenda. The protected areas strategy that has just been adopted for the next 10 years barely mentions rewilding (Ministère de la transition écologique 2021). It is only vaguely linked to other knowledge production objectives given to protected areas.

We cannot, therefore, rely solely on the argument of biodiversity conservation and its representatives to raise awareness in favor of the rewilding of abandoned landscapes. This socio-cultural evolution requires the effort of all actors to take a critical look at the place of these spaces in society. It is, therefore, as much a question of understanding the complex ecological systems of these spaces as of what they can provide in terms of services or benefits to humans that must be studied in depth. However, we should not underestimate the initiatives and the current positions in favor of rewilding. In France, it has been studied and advocated for many years by ecologists (Génot and Schnitzler 2013; Cochet and Durand 2018; Cochet and Kremer-Cochet 2020), foresters (Vallauri *et al.* 2010), many naturalists and increasingly by non-professionals in environmental protection, who propose a less technical and more sensitive view of nature. Personalities mobilize to advocate

---

2 See: https://artificialisation.biodiversitetousvivants.fr/cartofriches/enjeux-revitalisation-friches.
3 See: https://inpn.mnhn.fr/espace/protege/stats figures accessed March 15, 2021.

an uncontrolled nature[4], or to initiate concrete protection projects, like the botanist Francis Hallé, who proposed to create a primary forest without human intervention for 10 centuries[5]. More discreetly, many local initiatives like the PRELE network or the ASPAS (mentioned in the first part of this chapter) work in the territories to give a place to rewilding.

It is, therefore, essentially a change of scale that must be encouraged. First, a change of geographical scale. Today, at the scale of France, rewilding areas represent mere isolated dots in the landscape. Yet, the tensions generated by local initiatives (such as the creation of the 490-hectare Wild Life Reserve in the Vercors) show how strong the fears linked to wilderness are and how the idea of leaving a space to itself triggers traumatic echoes. As Morizot (2019) points out, "laissez faire" is synonymous for many people (the "decision-makers" in the forefront) with "rural desertification, loss of control over the territory, de-inscription of the human, invasion of wildness – and everyone wants to fight against it, without really knowing what they are talking about. Because the idea of a world left to itself terrifies", and in France, perhaps more than anywhere else, where agriculture is a national economic issue (France is the first European country in terms of agriculture) (Reynaud 2018) and extractive activities such as hunting are largely supported by the state (France is the European country that authorizes the largest number of bird species for hunting, European Commission 2009).

It is also a change of thematic and disciplinary scale that we must hope for. The change in the way we look at these areas will probably be concomitant with the growing awareness of their existence and value. The scientific study of their contribution to the mitigation of the effects of climate change, their resilience capacity, their effect on the well-being of local populations, etc. would certainly allow us to view them in a more positive light and accept them more easily in our landscapes. In this respect, geography, with its interdisciplinary approach, seems particularly well-suited to the study of rewilding, both in its spatio-temporal dimensions and in terms of the complex relationship that humans have with their environment. While the main challenge is to leave physical space in our landscape for rewilding, this is matched by the need for us to find space in our imagination to accept and embrace wild nature.

---

4 Opinion column of a collective of personalities published in the newspaper *Le Monde-Opinions* on December 8, 2020. "Arrêtons de vouloir maîtriser et exploiter la totalité des espaces et des écosystèmes".

5 Opinion column by Francis Hallé published in the newspaper le Monde-Opinion on October 7, 2019. "Il y a urgence à reconstruire de grandes forêts primaires".

6.5. Conclusion

Rewilding is a concept that includes at least two distinct forms. Active rewilding aims to restore a character and functioning similar to what it could have had before anthropic domination by human intervention. This form was dealt with in Chapter 5. But another form of rewilding can be distinguished, that which appears spontaneously following the abandonment or at least the relaxation of anthropogenic pressure. It can occur more or less voluntarily and more or less abruptly. The ecosystems and landscapes that result from this spontaneous rewilding are not those of the past and have the characteristic of having intact self-sustaining ecological processes (without human hindrance). Depending on the angle from which these spaces are analyzed, they can be described as a "novel ecosystem" if we are interested in their ecological functioning, of a feral nature if the analysis is more historical, as abandoned lands for those preoccupied with human uses, etc.

In terms of the baseline notion discussed in this book, spontaneous rewilding can be said to be the opposite of active rewilding in that it ignores any reference to a past state. Whether intentional or not, spontaneous rewilding can be compared to other spontaneous dynamics that exist or may have existed elsewhere in space or time. In this sense, spontaneous rewilding can be seen as a return to a past baseline of intact natural processes, but producing singular environments that are not comparable to historical states.

Areas of land abandonment are probably the most important areas of rewilding in terms of surface in France and perhaps even in Europe. These areas are relatively undocumented because they are located at the margins of the various sectors of interest. Neither agricultural areas, nor urban areas, nor even natural habitats, areas of land abandonment have long escaped the radar of planners. However, these forgotten areas may not be forgotten for much longer. Various initiatives aim to better identify them so that they can be used as land reserves for future development projects. Since they are not counted, these spaces could help balance the "zero net artificialization" objective of the 2018 biodiversity plan. However, the positive effect on biodiversity of promoting the transformation of these open spaces over agricultural spaces is not obvious and at least deserves questions to be asked collectively.

On the scale of metropolitan France, the areas of land abandonment represent at least 1,000 km^2 (see Figure 6.4). This figure – certainly very conservative – only concerns areas where agricultural use was abandoned at least 10 years ago, because the calculation method implies that the ecological succession (from an agricultural herbaceous stage to an intermediate shrub stage) is visible on aerial photographs. For comparison, the naturally evolving terrestrial protected areas (strict biological reserves and strict national park reserves) do not represent a larger area. Obviously, the ecological and especially patrimonial interest of these abandoned areas is not comparable to the old, mature forests that have been free for centuries. However, today's abandoned lands could very well become tomorrow's vast wild forests. If we collectively decide to develop and conserve a network of naturally evolving environments, then these abandoned areas are a particularly, we may even say uniquely, well-suited candidate. They have enjoyed a spontaneous dynamic for at least 10 years, are already at a shrub stage and have escaped certain economic imperatives. For this, we will have to change our perspective in order to see not only a short-term "land supply", but also, and perhaps above all, a long-term supply of naturalness in these spaces.

7

Geoprospective: Looking for Potential Scenarios

7.1. Introduction

The future is, by definition, uncertain. While this can frighten, make us skeptical or doubtful, another attitude can be anticipation, which results in defining various adaptation strategies. For example, when buying a property, everyone has information about the risks (floods, landslides, technological risks) linked to the property's location. A lack of anticipation consists of not considering a risk of flooding, for example. A first strategy could be to accept the risk and adjust one's lifestyle (upstairs only) or the architecture of the property (raise it on stilts). A second strategy could be to refuse this location and look for another property outside a flood zone defined in a flood risk prevention plan (PPRI[1]). A last strategy could be to go beyond the geographical perimeter of the current regulations (look for an even more remote area, another locality or region) on the assumption that the current regulation defining PPRI is insufficient with regard to climate change.

While this example illustrates an individual choice, societal demand has evolved and the growing awareness of the concept of sustainable development since the 1970s (Meadows 1972) has made it possible to integrate this anticipation into current land-use planning tools (*Schéma d'Aménagement et de Gestion de l'Eau* (SAGE), *Schéma de Cohérence Territoriaux* (ScoT), etc.) because of their long-term scope.

Anticipation through complex thinking (Morin 1990), based on the principle of a future made up of several possible futures (or "futuribles" according to de Jouvenel

---

Chapter written by Thomas HOUET.

1 In France, these are called PPRIs, the acronym standing for *Plan de Prévention du Risque Inondation.*

(1964)), has thus contributed to the emergence and generalization of the *prospective*. *Prospective*, which can be defined as an approach (Godet 1977, 1986) or a position (Berger 1958), including foresight studies, which "consists of gathering elements of appreciation, whether or not quantified, concerning the future, the aim of which is to make it possible to make decisions which will enable the said future to be more in line with our preferences than if there had not been this prospective insight" (Hatem *et al.* 1993). While this definition refers to a baseline (successfully conforming to something), the plurality of possible futures gives it an eminently polysemic character.

And then, geography got involved… both through this book, for example, which questions the baseline notion, and through approaches and methods designed to take into account the spatial dimension in the construction of these visions of the future. Territorial foresight (Delamarre 2002; Loinger 2004), geoprospective (Voiron 2006; Emsellem *et al.* 2012; Gourmelon *et al.* 2012) and prospective modeling (Houet 2006; Houet *et al.* 2011) are all concepts and methods that have contributed to refining, or even transposing, this baseline notion in geography or ecology (Albert *et al.* 2020), but also to expanding it.

This chapter attempts to shed some light, albeit probably non-exhaustive, on the various meanings that the notion of a baseline can have in geoprospective. While the particle "geo" is the prerogative of geographers, it has no other ambition than to be able to integrate the spatial dimension in all its complexity, in prospective. Indeed, while the multidisciplinary character is intrinsic to prospective, *i.e.*, taking into account social, economic, historical, cultural, demographic, political, etc. dimensions, the geographical dimension was only used until the 2000s as a cartographic medium to represent future scenarios. Geoprospective is distinguished by the integration of spatial processes resulting from nature–society interactions (neighborhood, inertia, multi-scalar relationships) to refine the location of future land use and/or landscape changes. Even if spatial simulation models are used to produce scenarios, a common approach that is used to explore a wide variety of possible futures consists of implementing, firstly, participatory methods intended for the construction of narratives and which will, secondly, feed spatial simulation models that are intended to produce cartographic representations of them (Houet *et al.* 2016a, 2017). As the spatial dimension can be integrated at all stages of scenario construction (Houet *et al.* 2010), this chapter aims to illustrate, through examples, the various meanings of the notion of a baseline in geoprospective.

The construction of prospective scenarios is based on three stages (Godet 1986): (1) the construction phase for the knowledge base, which is necessary and shared for the construction of scenarios (Poux 2003), (2) the scenarios themselves, made up of narratives and images of the future and (3) the action phase, during which quantified elements of appreciation are provided in order to be able to make decisions and define a strategy. Four meanings of the notion of a baseline can be defined in geoprospective:

– The first occurs during the construction phase of the knowledge base, which must be shared by all the people participating in the prospective exercise. Thus, even if the points of view and objectives diverge, the knowledge base constitutes a common baseline.

– The second meaning concerns the scenario construction phase. It refers to the notion of validation, in particular, of the methods used for their construction, in order to provide supporting elements that are intended to reinforce the confidence that the different players will ultimately have in these scenarios.

– The last two are part of the action phase: they can be used to evaluate, in an exploratory way, the impacts of these scenarios in relation to the initial situation (the current one), but also to define an objective to be reached in order to define the actions to be implemented.

7.2. The baseline as a shared and objective knowledge base

The definition of a shared knowledge base intrinsically refers to the first part of this book, "defining baselines". In prospective, it is a matter of defining the current state, the past dynamics, the systemic functioning (driving and dependent variables and their causal relationships), the projects and the seeds of change in the system being studied. As soon as this system is assimilated to a territory, the spatial dimension can be integrated via multiple geographic methods: remote sensing, geostatistics, participatory methods (social surveys or participatory mapping, for example), modeling of complex systems, etc.

Hatem (1993), Piveteau (1995) and Poux (2003) have identified one of the main limitations of a prospective exercise: the tyranny of received ideas. Very often, participants have fixed ideas about certain landscape evolutions and their explanatory factors as a result of their personal experiences and perceptions. For example, while the disappearance of mountain agropastoral landscapes is well-known, the areas concerned and the rates of progression of encroachment and reforestation may be largely underestimated (Houet *et al.* 2012). Indeed, between 1990 and 2006, the Corine Land Cover database detected almost no change in land uses in the Pyrenees, whereas, using higher spatial and temporal resolution remote sensing data, clear changes were visible between the eastern and western Pyrenees. Similarly, using aerial photographs, it was shown that the Vicdessos valley has, since the 2000s, experienced a process of encroachment and afforestation that is as rapid as the slopes or valley bottoms during their peak rate after World War II (in the order of 50–60 ha/year) (Figure 7.1).

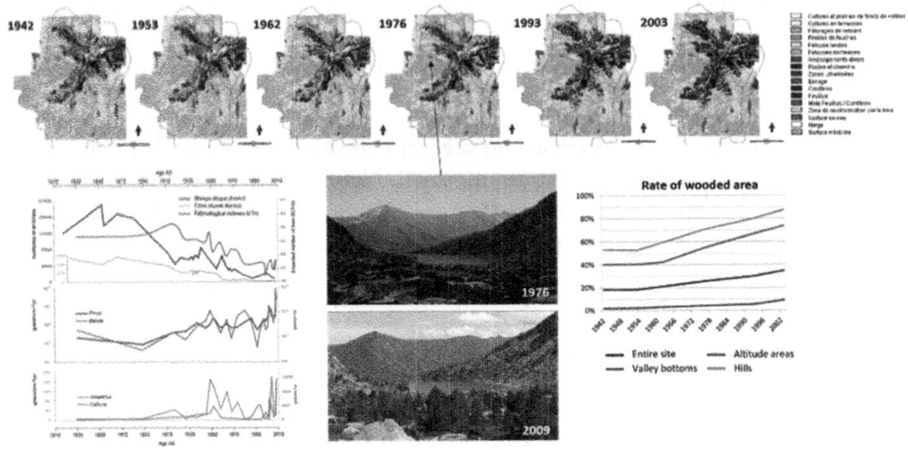

Figure 7.1. *Illustration of the contribution of multidisciplinary approaches to establishing an objective and shared knowledge base using perceptual elements (landscape photographs) and figures (maps and graphs). The example is that of the Haut-Vicdessos valley (and the Bassiès valley) and the disappearance of the landscape (from Galop* et al. *2011 and Houet* et al. *2012). For a color version of this figure, see www.iste.co.uk/godet/baselines.zip*

These received ideas may also have a time frame that does not match the time horizon of the prospective exercise. This may cause bias in the perception of certain practices. For example, slash-and-burn is now considered a cultural practice that is part of the pastoral heritage of the Pyrenees, and should be given special attention in the future. Galop *et al.* (2011) were able to demonstrate, using palynological data, that the appearance of mountain fires came later than the dynamics of grazing in the Vicdessos valley. In the 1930s and 1950s, slash-and-burn appeared to be an adaptive practice to fight against encroachment following the decline in pastoral activity that began at the end of the 19th century. This knowledge makes it possible to envisage the non-maintenance of these practices in certain scenarios, since they have not always existed. Similarly, the oral knowledge shared during participatory meetings on the former uses (pastoral pressure) of these summer pastures also made it possible to consider scenarios where certain summer pastures are abandoned in exchange for increased pastoral pressure on others (Houet *et al.* 2017) (Figure 7.1). While the divergence of viewpoints is essential in order to consider highly contrasting scenarios (to fight or not fight against land abandonment), it must reflect the stakes for the local actors involved and not be dismissed by the tyranny of received ideas. Thus, this knowledge base that is necessary for the construction of scenarios constitutes a tangible and shared baseline.

7.3. The baseline as a way to improve confidence in scenarios

The knowledge provided by scenarios must be used with confidence by local actors in order to establish their action plan. Although prospective does not aim to predict, it is the plurality of possible futures that will make it possible to define management and land-use strategies that are not as harmful as they would have been had there had not been this insight into the future. But then, how can scenarios be verified and validated, since the future cannot be validated, on the one hand, and since, on the other, the paradigm of the prospective is opposed to that of prediction? This dilemma is particularly marked when modeling tools are used to simulate prospective scenarios.

This confidence is therefore focused on demonstrating that the methods and tools mobilized are working "as they should" or as predicted by the scenarios (as a set of parameters) and not on validating the results as such. In order to validate a prospective exercise, Piveteau (1995) proposes checking whether four principles are verified: relevance, transparency, plausibility and consistency.

Transposed to geoprospective, they question the use of (spatial) models in prospective (Houet 2015), whether they are cellular automata, multi-agent systems or other modeling platforms. These tools are more and more commonly used to simulate scenarios of land use and land cover changes. "Validation" is becoming more complicated and can no longer be summarized in a single performance indicator (Figure 7.2)

– The relevance is the need to justify the choice of a model by an analysis of the literature and/or a comparative study of existing models. The aim is to demonstrate its ability to simulate landscape changes that make sense for the territory or the issues studied (*e.g.*, crop rotations, urban sprawl, etc.) in terms of both their spatial representation and their temporal dynamics, and even the underlying decision-making processes.

– Respect for the principle of transparency is demonstrated by the ability of the user to understand and explain the functioning of the model used. The aim is to avoid the "black box" effect which, if presented in this way to local actors and other participants in the prospective exercise, can sow doubt, mistrust or even constitute an excuse to reject all the results.

– Plausibility is part of a "classic" model validation logic, where the predictive power of the model is evaluated through specific indicators, either used in isolation or combined (Paegelow *et al.* 2014). It is then a question of carrying out a simulation on a past period and comparing it with an observed situation.

– Consistency must allow for demonstrating that the model is able to simulate the processes or trajectories that are envisaged by the scenarios. Unit, integration or sensitivity tests can thus demonstrate that new spatio-temporal processes can be simulated, in coherence with other existing or new processes.

Compliance with these four principles is fundamental because, while a model can perfectly meet the plausibility principle by simulating a past trajectory perfectly, its calibration and the parameters it uses can make it unsuitable for simulating scenarios that are at odds with past trends (Houet *et al.* 2016b). Moreover, let us also recall that the plausibility of simulations depends on the plausibility of scenarios, which must be checked beforehand. A model, once validated, can, in any case, plausibly simulate a completely inconsistent scenario.

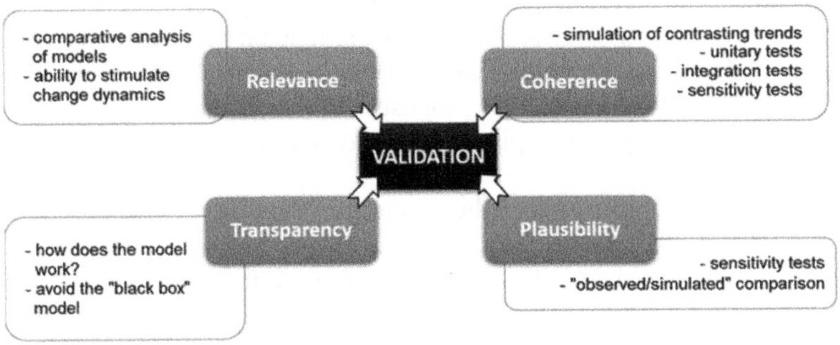

Figure 7.2. *Illustration of the methods that can be used in modeling to respect the four principles of validation of a prospective approach*

7.4. The baseline for the exploratory evaluation

Prospective is intrinsically based on a plurality of possible futures. Knowledge about the future therefore comes from the multitude of situations, be they desired or not, and from their comparison. The analysis of future changes in land use and land cover patterns allows us to evaluate the impacts (positive or negative) on the environment or ecosystem services, for example. The comparison of future states is therefore made in relation to a baseline situation, often assimilated either to the current situation or to another scenario, called the baseline scenario.

A classical approach therefore amounts to comparing the future image resulting from a trend scenario (of the “what if nothing changes” type, or integrating future elements whose occurrence is certain) or not, with the image of the present situation. This allows us to, for instance, quantitatively assess the evolution of landscape closure in the mountains (Houet *et al.* 2015; Vacquié *et al.* 2015) (see examples in Figure 7.3). It is then possible to assess the impacts of these changes on hydrology (Szczypta *et al.* 2015) or even on gravity risks (Bernardie *et al.* 2021). The advantage of using modeling tools for this assessment is that it is possible to distinguish between the part that is attributed to climate change, on the one hand, and the part that is linked to anthropogenic activities (or lack of activities) causing a change in the landscape, on the other.

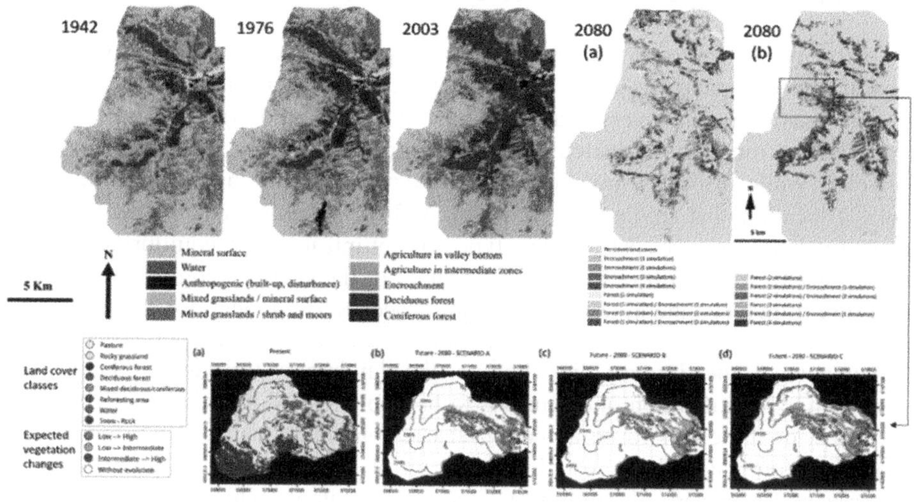

Figure 7.3. *Examples of exploratory scenarios of environmental closure in the Haut-Vicdessos Valley in 2080 (top) for two pasture management strategies: a) maintaining low pastoral pressure everywhere and b) abandoning the least favorable summer pastures (from Houet* et al. *2015), and their variation at a more detailed scale for levels of (A) strong, (B) medium or (C) zero pastoral pressure in Bassiès (source: Sczcypta* et al. *2015). For a color version of this figure, see www.iste.co.uk/godet/baselines.zip*

Whatever the baseline considered, whether the present situation or a future situation, this approach is fundamentally exploratory, aiming at evaluating what would happen/what would be gained if…

7.5. The baseline as an objective to be reached

As soon as the scenarios are normative, *i.e.*, starting from a future to be achieved (or avoided), the future becomes an objective, a baseline to be achieved. Any deviation from the desired trajectory can then constitute a warning signal of the ineffectiveness of the policies at work. Nevertheless, we must also take into account the potential inertia of the socio-ecological system with respect to the effects of the measures at work, which may not be immediately perceptible. The latest IPCC scenarios fall into this category of scenarios: the description of the "representative concentration pathways" (RCP) defines four trajectories of greenhouse gas emissions and concentrations resulting from a multitude of possible socio-economic and technological scenarios. However, regardless of macroeconomic, demographic or other assumptions, if the situation is close to a pessimistic trajectory (RCP 6.0 or 8.5), it will simply reflect the failure of regional, national or international policies.

This type of approach is very rare in geoprospective because the backward construction of spatio-temporal landscape trajectories from a future goal to the present situation remains very challenging. Brunner *et al.* (2016) propose an original and efficient multi-agent system modeling approach for this type of approach. However, while it clearly has the virtue of being available, this type of modeling seeks a location that minimizes land-use conflicts, resulting in an often "direct" trajectory between two dates. This type of optimization or equilibrium search method makes it difficult to model a bifurcation of the socio-ecological system (the trajectory from a neoliberal use model to an agroecological model, for example).

Sometimes, the interest lies not in the landscape trajectory, but in the retrospective identification of the means to be implemented in order to achieve this landscape, or the errors to be avoided (lack of interventionism). For example, the Rennes agglomeration has created the concept of "urban fields" in its *Schéma de Cohérence Territorial* (Territorial Coherence Scheme). This zoning makes it possible to protect agricultural land located in the immediate vicinity of Rennes and largely under pressure from urbanization, which has been made unbuildable in the local urban plans of the municipalities concerned. The goal is to preserve the landscape qualities, the ecological environments of interest and to support development by structuring commuting and linking recreational areas. This would not have been possible without having first identified the objective of implementing a concrete green belt and pursuing development as an "Archipelago City".

Whether it constitutes a monitoring indicator in relation to a "desired trajectory", or a means of defining the regulations or actions to be implemented in order to achieve a territorial objective, the future, in the normative sense, thus constitutes a particularly useful baseline for present action.

7.6. Conclusion

The notion of a baseline in geoprospective thus has a multitude of possible meanings, summarized in Figure 7.4 and taking up the three meanings defined in the introduction of this chapter. Globally, we can distinguish between an "observed and shared" baseline (the construction of the knowledge base), a baseline that is "made explicit" by modeling and finally a "desired" baseline that is linked to the future. Most of these notions refer to aspects of scientific, rigorous and objective methods. In this case, the baseline is not called into question, like the idea of the presence of the bear mentioned by the author in the general introduction to this work. It is essentially a starting point, both relative and subjective, concerning the senses, but which, once compared to the various possibilities that the future represents, brings a new, objective and quantified light.

In fact, he mentions three reasons that may call into question the idea of seeing a bear. Firstly, that it may be implausible and should not be considered. In this example, implausibility is only a human concept, or even a manipulation or a simple oversight. The knowledge of history is thus fundamental to contradict this first reason, because it is the means to fight against the tyranny of the received ideas, mandatory to any prospective.

The second reason concerns the position between science and the senses. It is a state of affairs, I myself have long felt a bit torn: on the one hand, divided between the duty of preserving that a scientist must have when they lead and animate a participatory foresight exercise, where only the opinion and the desires of the participants must be taken into account in order to hope for a collective strategy (they would certainly not adhere to the opinion of a scholar who is foreign to their territory proclaiming to them what is good or bad); and, on the other hand, that of the scientist activist who must incite a more sustainable development or to alert on the urgency of acting… Nevertheless, being aware of such a pathology allows us to take a step back, to avoid being the reason, as a scientist, for a possible failure of a prospective. It is clear that, on the part of the participants, this type of exercise has a kind of medicinal virtue: when we talk about the future, people who no longer talk to each other start to speak again. Current tensions are, for the duration of a meeting, put aside for a while. The exploration of the future (carried out with rigor and method) destroys positions of principle and divergence because the future intrinsically evokes a common good and a collective interest. This is what geoprospective is all about: at the beginning, the opinions and points of view on the issue of a territory illustrate the divergences of the stakeholders; the scenarios allow for the emergence of ideas, of desirable or undesirable futures; finally, the knowledge on the future allows a convergence of interests and even the definition of a consensual strategy for the development and management of the space.

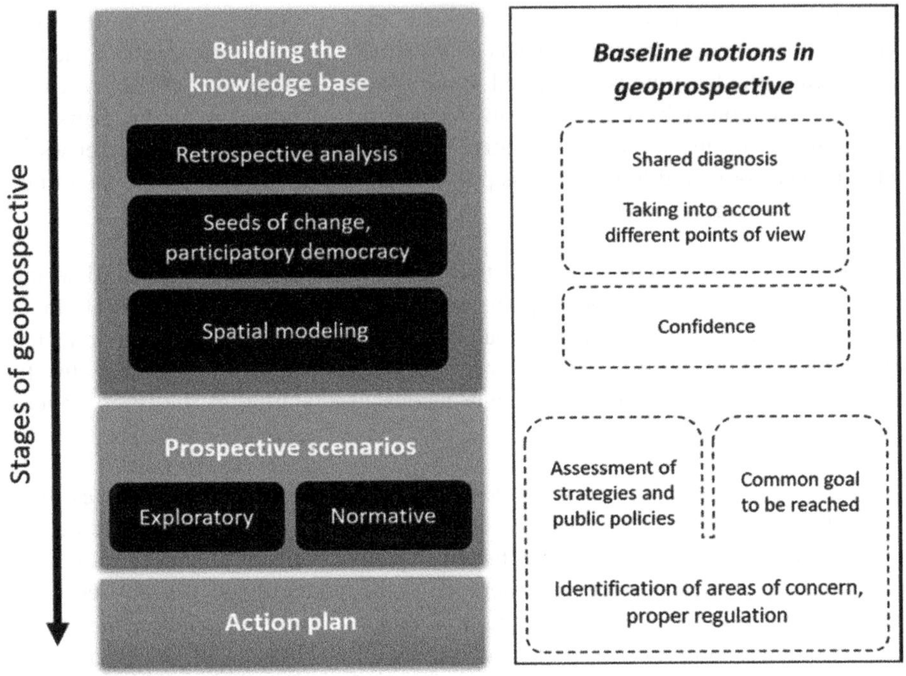

Figure 7.4. *The different variations of the notion of baselines in geoprospective according to the stages of construction of spatialized prospective scenarios*

Finally, the third reason concerns the Anthropocene itself, where human action would already be too significant and nature would no longer exist. Without getting into this philosophical or even ideological debate where Nature and Society are opposed by definition, but both deserve a place on this Earth, would it not be simpler to consider this as one scenario among others? And if the challenge consisted of making a prospective of the Anthropocene? …and then I realize. This has already been done by Meadows *et al.* (1972). So, what went wrong? Probably the social acceptance of this famous concept of sustainable development, the acceptance of the failure of a development model, the challenge of a carefree generation. Still, without this, the seeds of change would not have been sown. Today, the future generation is ready, but it is up to it to decide where it wants to go.

8

The Place of Ecological Knowledge in Policies for Ecological Neutrality: No Net Loss and Biodiversity Offsetting

8.1. Introduction

Today, it seems difficult to talk about biodiversity conservation without addressing the topic of biodiversity offsetting or compensation, and, more broadly, the mitigation hierarchy (also known as the Avoid–Reduce–Compensate or ERC (*Éviter–Reduire–Compenser*) sequence) within which it theoretically occupies the last place. This tool has, in recent decades, become one of the main levers of action in the field of biodiversity conservation (Levrel *et al.* 2018). The mitigation hierarchy aims to neutralize the ecological impacts of land-use planning on biodiversity according to a principle that obliges developers – whose projects, plans and development programs are subject to an environmental impact assessment[1] – to implement, as a priority, measures to avoid these impacts, then to reduce those that are deemed unavoidable and finally to compensate for those that remain qualified as "residual". In this last case, developers must compensate for the residual impacts of their developments by implementing ecological actions in order to counterbalance the ecological losses with equivalent gains, thus aiming for no net loss (NNL) of biodiversity. The notions of additionality and ecological equivalence are at the center of the application of biodiversity offsetting, and the questions relating to the definition of baselines are central to it.

Chapter written by Coralie CALVET.

1 In French environmental law (Article R122-2 of the Environmental Code), this mainly concerns linear infrastructures, business parks, quarries, ports, etc.

By addressing the main causes of biodiversity erosion, namely the destruction of habitats and ecosystems and their fragmentation due to land-use planning, this principle, which aims for ecological neutrality, then appears to be a pillar of the integration of biodiversity into land-use planning. It is thus becoming ubiquitous, whether in scientific and political arenas or among conservation actors "on the ground" (Calvet *et al.* 2015a). Biodiversity offsetting is a cross-cutting topic that now affects public decision-makers, planners, politicians, researchers and citizens, whether simple users or nature lovers.

Although biodiversity offsetting has only recently taken off, it has quickly become the subject of intense controversy involving scientific, practical and ethical considerations. The presupposition on which it is based, namely, the possible substitutability of ecological entities, raises concerns that refer to the way in which contemporary scientific ecology and social sciences consider questions relating to the choice of biodiversity to be preserved, to the methods and indicators to evaluate it and to the techniques available for replacing it. These questions are an opportunity, in turn, to study how the notions of baselines and ecological equivalence are grasped in science, which I propose to explore in this chapter.

Firstly, in order to provide an overview of the use and application of biodiversity offsetting, and, more broadly, of the mitigation hierarchy, I propose to summarize the regulatory developments and their translation into the concrete application of this tool at the international level, particularly with regard to the "equivalence issue" of biodiversity. In section 8.3, I will summarize the main problems that arise today in the application of the mitigation hierarchy in relation to the notion of a "baseline" to highlight the impasses between theory and practice. This scientific state of the art will lead me to mobilize the scientific knowledge produced in ecology, conservation biology and also in human and social sciences, given the plurality of issues surrounding the application of NNL policies. Finally, in a concluding section, I will reposition this tool in its spatio-temporal trajectory by focusing on the origins and conditions of the emergence of this policy. The aim will be to highlight the geopolitical context in which it appeared on an international scale. This historical retrospective seems essential for a better understanding of the choices made to apply this policy in our contemporary societies and their implications on the way we "practice" biodiversity conservation.

8.2. Global overview of the application of the mitigation hierarchy

8.2.1. *The success of NNL policies at the international scale: the political promise*

At every major meeting of world leaders in recent decades, governments have reiterated their commitment to "halting biodiversity loss"[2]. In the face of this ambitious goal, biodiversity offsetting is presented as having "the potential to achieve the broad biodiversity conservation goals set by governments, while offering solutions to land-use planning issues by overcoming conflicts between economic development and the public interest in biodiversity protection" (UNEP-CBD 2010, p. 1). The political stakes around biodiversity offsetting are clear: it carries the political promise of a possible reconciliation between the two often antagonistic issues that are economic development and biodiversity conservation (Boisvert *et al.* 2013).

Thus, the mitigation hierarchy, although present for decades in the laws of many countries, has reappeared on the international scene as the "spearhead" of policies aiming at no net loss of biodiversity, or NNL policies (*Ibid*). In a context where the demand for a reconciliation between "economy and biodiversity" is displayed as an international leitmotif[3], the success of biodiversity offsetting on a global scale has been swift since the 2010s (Calvet *et al.* 2015a). The NNL principle has since spread around the world, as highlighted by the GIBOP (Global Inventory of Biodiversity Offset Policies) database, which estimates that it is part of the public policies of more than 100 countries, 37 of which have introduced it into their regulations (Figure 8.1) (GIBOP 2019). The land footprint of biodiversity offsets is becoming increasingly prominent internationally. A recent study found that biodiversity offsets cover an area of over 150,000 km^2 across 37 countries (Bull and Strange 2018). At the scale of individual countries, this political momentum has been translated into regulatory developments specific to each country.

---

2 This commitment was first mentioned at the Conference of the Parties held in Nagoya in 2010, and then repeated in 2012 at the Earth Summit held in Rio de Janeiro, and again reiterated in 2020 at the 13th United Nations Conference on Biodiversity in Cancún. At the European level, the EU Biodiversity Strategy 2020 recalls this objective in relation to the 15th "Aichi 2020" target on ecosystem restoration, and aims to ensure "no net loss of biodiversity and ecosystem services" (EU No Net Loss initiative 2014).

3 As shown by the slogan of the Rio+20 Conference of the Parties with "A green economy for a planet under pressure" (https://sustainabledevelopment.un.org/index.php?page=view&type=400&nr=662&menu=1515), and the now well-known work of TEEB for "The Economics of Ecosystems & Biodiversity" (http://teebweb.org).

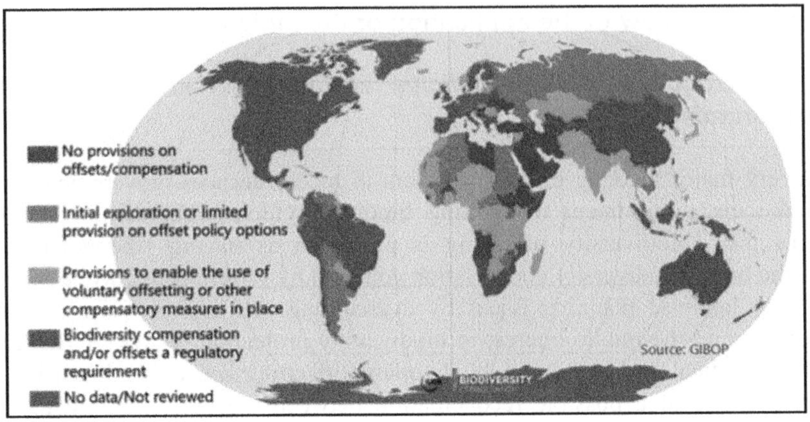

Figure 8.1. *Global mapping of the implementation of no net loss policies worldwide (source: GIBOP 2019). For a color version of this figure, see www.iste.co.uk/godet/baselines.zip*

8.2.2. *The application of the mitigation hierarchy in France: a recent regulatory evolution*

Implemented in a growing number of countries, the first mention of biodiversity offsetting, and, more broadly, of the mitigation hierarchy, the "Avoid–Reduce–Compensate" sequence, dates back to the 1970s. Integrated into the nature protection laws of many countries (France, Germany, the United States and Australia, to name the main ones), this sequence was not, however, applied during this period (Quétier *et al.* 2014). It was only in the 2000s that it started to be really mobilized, following numerous regulatory reinforcements. In France, for example, this application follows a series of decrees and orders aimed at specifying the regulatory terms (Levrel *et al.* 2018).

In the following, I have selected the main regulatory developments that have significantly impacted practice in relation to the mitigation hierarchy, and which are particularly enlightening in relation to the issues I am dealing with.

The most emblematic case concerns the decree of February 19, 2007, which sets the conditions for requests for derogations from the law against the destruction of protected species of wild fauna and flora. One of these conditions was the implementation of the mitigation hierarchy in the case of impacts on protected species. The consequences of this decree were, on the one hand, a surge in requests for derogations for the destruction of the habitats of protected species, and, on the other hand, a focus on the assessment of impacts and offsetting measures on protected species (Regnery *et al.* 2013).

The reform relating to impact studies for development projects, as set out in the decree of December 29, 2011, has also significantly modified practice in relation to the mitigation hierarchy. This modification mainly concerns the obligation of developers to carry out an initial assessment of the impacted site – a sine qua non condition for the introduction of the baseline notion in the mitigation hierarchy – and monitoring on offsetting sites (note that there is no obligation to carry out an initial assessment on sites used for biodiversity offsetting).

More recently, the changes concern the implementation of a "single environmental authorization" in 2017[4], which replaces the various independent authorization procedures to which a project could be subject, such as the authorization for protected species derogation, the authorization for land clearing, the authorization for the law on water and the Natura 2000 impact study. The purpose of this single authorization is to gather together the various procedures for examining a single development project that is subject to an environmental impact assessment study[5]. It is important to mention that this ordinance follows the government's desire to accelerate and simplify the examination of applications for development project authorizations[6]. Indeed, with the increase in procedures and the number of requests for derogations, authorization for development projects was becoming increasingly long and difficult to obtain, thus constituting an obstacle to land-use planning (Doussan 2018).

Finally, the most recent regulatory development for the mitigation hierarchy can be found in France in the implementing decrees of the law of August 8, 2016 for the recovery of biodiversity, nature and landscapes. This law makes major changes on several points. The mitigation hierarchy is displayed with the objective of achieving "no net loss of biodiversity". Offsetting measures must be evaluated and monitored in order to demonstrate "quantified" results concerning the achievement of ecological equivalence between the impacted and compensated sites.

The notion of ecological equivalence is central to this tool because it is at the heart of the biodiversity offsetting principle. The calculation of ecological equivalence implies making choices in terms of the methods and reference system for evaluating the state of the impacted and compensated sites. We will go into more details on these elements in the next section, which deals with the ecological issues involved in the application of biodiversity offsetting.

4 Executive Order No. 2017-80 and Decree Nos. 2017-81 and 2017-82 of January 26, 2017.

5 See the Annex to Article R. 122-2 of the French Environmental Code.

6 See the report available at: http://www.senat.fr/rap/r18-560/r18-5602.html. This report, written at the request of Jean-Marc Ayrault, then Prime Minister under the Sarkozy government, explicitly highlights the political desire to "reduce the influence of standards downstream of their production" and the need to move towards "a facilitating interpretation" of them.

Changes in regulatory requirements have led to the establishment of a major organization around the application of the mitigation hierarchy, involving a diversity of actors: project owners, naturalist consultancies, managers of natural areas, land specialists, actors from the agricultural world, regulator services, public decision-makers, scientists, etc. Thus, new arrangements for biodiversity conservation actions have emerged, contributing to the reorganization of the biodiversity conservation and management systems already present (Guillet and Sémal 2018). New roles have been entrusted to actors in the field of conservation, and new actors have entered the conservation landscape, particularly from the private sector, with their own strategies. These developments bring new power relationships, new sources of competition and potential conflicts within conservation actions (Gordon *et al.* 2015; Guillet and Semal 2018). However, depending on the different countries, a great diversity of practices and organizational responses exists, resulting in a significant disparity in methodologies and consideration of ecological components and processes.

8.2.3. *The implementation of NNL policies: a high degree of heterogeneity at the international level*

At the global level, biodiversity offsetting is applied either in a regulatory way, for 37 countries, 10 of which impose it as part of the mitigation hierarchy, or in a voluntary way, for 64 countries (Figure 8.1). The latter may be motivated, for example, by obtaining funding from international financial institutions. Depending on the regulations of the countries, the scope and modalities of the application of biodiversity offsetting differ.

As examples, I will present the regulatory frameworks and practices of four countries that have strongly contributed to the establishment of norms and standards in the implementation of NNL policies at the international level, thus feeding numerous scientific works: France, Germany, Australia and the United States.

In France, the mitigation hierarchy must be applied within the framework of environmental assessment procedures for the impacts of development projects, theoretically, on all components of biodiversity. This includes protected species, aquatic environments and wetlands, Natura 2000 sites, wooded areas and ecological continuity. In practice, protected species account for almost all of the biodiversity offsets implemented, which is not without problems of the effectiveness of public action with respect to biodiversity conservation issues (Regnery *et al.* 2013; Guillet and Sémal 2018).

In Germany, the mitigation hierarchy addresses all impacts on all environments (including landscape impacts), and it applies both under the planning code and the Nature Protection Act (Bas *et al.* 2019).

In the United States, pioneers in the regulatory introduction of the NNL policies and the mitigation hierarchy, the focus is more on wetlands, whose regulations, established by the Clean Water Act, introduced offsetting obligations and the mitigation bank system (Hough and Robertson 2009).

In Australia, land clearing of native vegetation, such as forests composed of endemic species, is prominent in offsetting practice, but this remains highly jurisdictional. Australia has also pioneered the implementation of biobanking, a form of offset banking, within the jurisdiction of New South Wales (McKenney and Kiesecker 2010).

Depending on the choice of practices established in these different countries, the requirements for ecological equivalence are very different and of varying levels of strictness and restrictiveness. For example, in France, ecological equivalence must be strict (*i.e.*, forest losses must be replaced by forest gains), and in-kind (*i.e.*, equivalence based on biophysical approaches), called, respectively, "like for like" and "in-kind" offsets in international terminology. This implies that the gains brought by biodiversity offsets must be evaluated and sized in relation to the identified ecological losses. Offsetting is mainly carried out within the framework of a "case-by-case" approach, also called an "on-demand" approach, in which each ecological gain of the offsetting measures is sized in relation to a loss identified in a project. Since the 2016 Law for Biodiversity, the French government has proposed the use of the biodiversity offsetting bank system, called the "natural offset site". Within this system, ecological gains are anticipated and implemented by a third-party operator upstream of future impacts of developers with a view to pooling several ecological actions within a single larger scale. This system is struggling to emerge in France, mainly because of the mismatch between the institutional context and conditions for its effectiveness (Calvet *et al.* 2015b). For example, offsets carried out within the framework of the banking system are more expensive than those generally practiced in "case-by-case" systems, due to the scale of the ecological restoration actions carried out (over hundreds of hectares and with management planned for a minimum of 30 years). However, without a preference of the investigating authorities for one offset over another, these actions (which are supposedly more relevant from an ecological point of view) find themselves in competition with cheap offsets, for instance, those carried out simply by acquiring and preserving the already existing natural sites in order to avoid future degradation. In the United States, the regulator has set up a preference system to avoid this, in the case where offset banks propose offsets in the same geographical area that are more efficient ecologically.

In Germany, a more flexible equivalence is preferred in the form of "eco-points", which are then called "out-of-kind". In the latter, equivalence can then be assessed between different biodiversity components, with Germany also considering esthetic and landscape components, soil, air and climate in the assessment of development impacts (Bas *et al.* 2019). Within this offsetting system, a number of so-called eco-points is assigned to each type of land use (*e.g.*, a road is worth 0.5 points, a meadow 2 points, a pine hedge 1 point, etc.) based on the issues identified within the 13 biogeographic zones that make up the federal territory. Within the framework of development projects, the contracting authorities convert the impacts generated by their project into eco-points which they must then compensate by implementing ecological restoration actions generating the same number of eco-points (not necessarily targeting the same types of biodiversity components as those impacted). They can also use forms of offset banks, called *Ökokonto* for "Eco-account", which offer these eco-points. It is important to note that this form of offsetting, which does not provide for ecological equivalence between impacts and offsets, does not apply to protected species and habitats, but only to other components of nature, which are not currently compensated for in France. German laws are stricter on the possibilities of derogating from the protection of species resulting from European directives. Thus, in the case of impacts on protected species, the ecological equivalence must be very strong and the developers must clearly demonstrate the impossibility of avoiding and reducing their impacts.

In line with the ecological approaches developed in systematic conservation planning (Margules and Pressey 2000), the scientific literature highlights that the search for a more flexible ecological equivalence for certain components of nature, such as common biodiversity, may have the advantage of increasing the possible benefits of biodiversity conservation at larger territorial scales (such as landscapes and regions) by prioritizing the most relevant ecological actions to be carried out on an area (Gonçalves *et al.* 2015). However, this raises questions of which form of biodiversity to "prioritize" over another, which will be "sacrificed" (Brownlie *et al.* 2013).

In the United States, as in Australia, equivalence is generally like for like and in-kind. Mitigation bank systems are highly developed in these two countries, concentrating the majority of the biodiversity offsets performed (Bull and Strange 2018). These countries also offer another form of offsetting, which is financial. In this framework, no ecological equivalence is sought between losses and gains, but a simple financial transaction made, whose amount is evaluated according to the type and level of impacts. This financial compensation is then paid to biodiversity conservation organizations or associations. The latter collect funds from several different development projects in order to pool them and use them for larger-scale conservation actions on given territories. There is therefore no "ecological equivalence" between biodiversity gains and losses. I do not deal specifically with this type of offsets in this chapter, because it concerns only a minority of practices, and it does not involve the same issues in relation to the assessment of biodiversity and the NNL target.

Thus, in the face of this heterogeneity of practices, different methods for assessing ecological equivalence have been developed around the world (Bezombes *et al.* 2018). All the same, these methods have a common basis: they involve the calculation of ecological losses and gains via the application of approaches of varying degrees of complexity and scientific rigor and mathematical formulas. However, depending on the approaches chosen, the reference states of the impacted and compensated sites will be considered to varying degrees.

8.2.4. *Methods for calculating the ecological equivalence*

The international scientific literature gathers the methods for calculating ecological equivalence under the term EAM, for equivalence assessment methods (Bezombes *et al.* 2017), whose acronym we will keep in the rest of this chapter. Each country develops its own EAM, based on the practices mentioned above, drawing on each other to different degrees. For example, the American UMAM (Uniform Mitigation Assessment Method), developed to assess impacts on wetlands, has inspired many other methods, notably in France with the MERCIe method (*Méthode d'Évaluation Rapide de la Compensation des Impacts Ecologiques*). In general, the heterogeneous practices both within and between countries do not favor the emergence of a unanimously recognized standardized method. For example, in the United States, 24 EAMs were identified in 2015, compared to 13 in Germany and about 10 in France (Pioch *et al.* 2015; Méchin 2020). These methods are called rapid assessment methods (RAMs), because their objective is both to propose an assessment of ecological equivalence over a given area and, above all, to propose methods that are easy and quick to implement (within a time frame corresponding to one day) (Méchin 2020).

Measuring losses and gains involves choosing metrics and indicators and expressing them in a comparable unit. Not all EAMs are constructed according to the same definition of ecological equivalence. This results in differences in the calculation methods chosen. These choices are particularly crucial because they directly influence the type and level of biodiversity that will be considered in the calculation of ecological losses and gains, as well as whether or not ecosystem states will be taken into account in their assessment. Practice has evolved in recent years in the calculation of offsets from fairly simple methods based on area ratios (hectare/habitat) to more complex methods taking into account more variables and indicators (ecological uncertainty, temporary loss, ecological functionality, etc.)

(Boileau *et al.* 2022). The quest to take into account the ecological functionality[7] of environments is gradually leading some consultancies to develop tools for modeling ecological continuities, but this is still not widespread in the current practice (Calvet *et al.* 2019a; Tarabon *et al.* 2020).

Faced with the rapid development of a multitude of sometimes very heterogeneous methods, numerous working groups (WGs) have been set up at different territorial levels, ranging from the very local (*e.g.*, authorities and regional biodiversity agencies) to the national level (with the Ministry of Ecology and the French Office for Biodiversity), and at the international level (within the European strategies for biodiversity). From these WGs, different guides or proposals have been developed to support territorial actors in sizing biodiversity offsets. In her thesis, Agnès Méchin counted no less than 42 guides and reports, and 13 conferences held in France on this issue between 2010 and 2020 (Méchin 2020). All these institutional dynamics reveal the importance given to this question of dimensioning ecological equivalence in the implementation of NNL policies.

What is particularly interesting to note is that, depending on the forum, the approaches and methods put forward are not the same, thus influencing the ecological knowledge mobilized. For example, the sizing methods proposed by the French Ministry of Ecology group are mainly based on qualitative (such as multi-criteria analysis and expert opinion) or mixed (qualitative and quantitative) approaches, whereas those issued by the OFB forum tend to favor quantitative approaches in order to limit the place of expert opinion[8].

In the recent OFB report (OFB 2020), three broad categories of EAM methods were proposed with the following main characteristics:

(1) Methods based on a *minimum ratio* approach: this is a purely quantitative approach in which the impacts on the environment are translated into metrics (surface area, linear area or volume) and then multiplied by a pre-established minimum ratio in order to calculate the offsetting needs. For example, depending on the species impacted, the ratio can range from 1 to 4, *i.e.*, for every 1 hectare impacted, there must be 1 to 4 hectares to compensate. In practice, these ratios are established by the naturalist consultants who conduct the regulatory impact study files for the projects on behalf of the developers, according to their own methods (Dauguet 2018; Méchin 2020).

7 As defined by Noss (1990).

8 For a detailed description of these different approaches and examples of application, see the OFB report (2020) on pages 21–32, available at: https://professionnels.ofb.fr/fr/doc-comprendre-agir/dimensionnement-ex-ante-compensation-atteintes-biodiversite-etat-lart-approches.

(2) Methods based on a *weighted* equivalence approach: these methods estimate biodiversity losses and gains separately by weighting the metrics by coefficients of "losses" (for those impacted) and "gains" (for those to be compensated) before verifying their equivalence. The latter is evaluated on the basis of a combination of criteria that can be predefined using classes (low/medium/high) or typologies of operation types and ecological engineering works. These methods take into account the description of the initial state of the environment, their conservation issues at stake, the nature and intensity of the project's impacts, the ecological added value generated by the offsetting measures and other parameters to carrying out the ecological actions.

(3) Approaches based on equivalence *between differences in environmental status of ecosystems before and after impacts or biodiversity offsets.* This approach quantifies ecological equivalence by assessing biodiversity losses and gains on the two ecosystems separately (the impacted one and the compensated one). It is mostly focused on the assessment of the state of the ecosystems with regard to different components (physical, chemical, biological, etc.). They are then supposed to take into account the ecological functionality of ecosystems. To do this, the methods aim to estimate the potential evolution and trajectory of the ecosystems after the project, on the one hand, and after the implementation of biodiversity offsets, on the other. These approaches mainly use predictive models and local ecological data to do so. The calculation of gaps (between losses and gains) is then used to size the biodiversity offsets required. The trajectory estimates can then be used to monitor the future ecological outcomes as "baselines".

In Figure 8.2, I present an example of the application of the methods based on "*differences in environmental status*" in a large offsetting program (involving more than 1,000 hectares of land) for ecological impacts on Little Bustards (*Tetrax tetrax*) habitats. These impacts were caused by the construction of a new 80 km railway line in Southern France. In this context, a naturalist consultancy has developed an ad hoc method based on "*Compensatory Units* (CU)" in order to size the ecological losses and gains. The calculation of ecological losses and gains is carried out by evaluating the difference between the state of ecosystems *before/after* impact or offsetting (depending on whether we are talking about the impacted area or the area that will be used for biodiversity offsetting).

Figure 8.2 illustrates the calculation of ecological gains (following the implementation of offset measures). Depending on the land-use history of the offset plots, the difference between the state *before* (Part A of the figure) and *after* (Part B) the implementation of the offset actions will not have the same *value*. For example, as shown in Figure 8.2, the change from orchard (Part A) to grassland (Part B) accounts for 2.5 CUs per hectare, while the change from cereal crop (Part A) to Mediterranean grassland (Part B) accounts for only 1 CU per hectare. The objective of these offsetting actions is the creation of habitats that are favorable to Little Bustards. On the basis of this approach, in contrast to the area-based approaches, the CU method takes into account a level of effort in the offsetting that does not give the same value to each hectare. Thus, in the context of this project, the developer's total offsetting debt has been estimated at 3,279 CUs, to be maintained over 25 years (which corresponds to approximately 1,100 hectares of offsetting compared to approximately 2,000 hectares impacted). These approaches explicitly aim to limit the land holdings of the offsetting measures and their local social and economic consequences (*e.g.*, increase in the price of land, competition between users, etc.)[9].

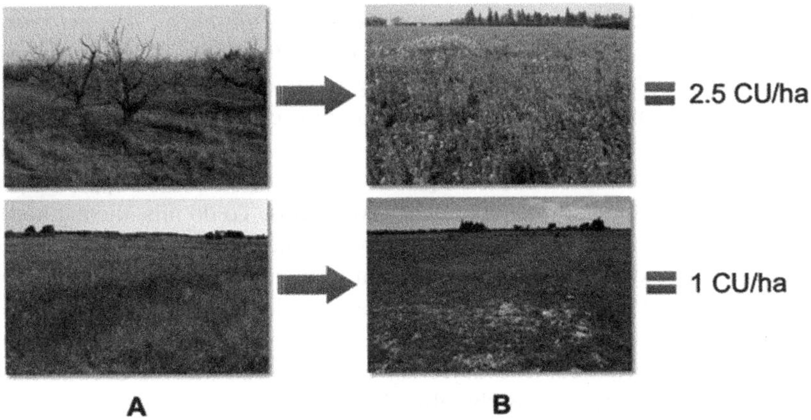

Figure 8.2. *Example of the application of an ecological equivalence assessment method based on differences in environmental status before (A) and after (B) the implementation of biodiversity offsets*

EXPLANATION OF FIGURE 8.2.– This approach emphasizes the notion of "quality" and "level of effort" of the offsetting actions carried out on ecosystems. Thus, the same hectare does not account for the same number of compensation units (CUs) depending on the difference between the state of the ecosystem *before* offsetting and that *after* offsetting. For instance, the change from orchard (at the top of Part A) to grassland (at the top of Part B) accounts for 2.5 CUs per hectare, while the change from cereal crop (at the bottom of Part A) to Mediterranean grassland (at the bottom of Part B) accounts for only 1 CU per hectare. The calculations are based on levels of "favorability" of the ecosystem for targeted species, in this case, the Little Bustard, as established by the naturalist consultancy, here Biotope.

9 For more information on the study of this project, see Calvet *et al.* (2019b), and on the CU method developed by Biotope, see Quétier *et al.* (2015).

Only the methods by weighting and differences in environmental status take into account a *baseline* of the environment. However, although these methods seem to be the most promising from an ecological point of view, they remain very marginal in their concrete use. Indeed, the study conducted by the OFB (2020) reveals that out of a sample of 135 projects evaluated, only half of them used a sizing method. Among these projects, the majority of the methods used was based on "ratio" approaches that estimated only the losses, without taking into account the ecological status of the environment in both the impacted and compensated areas (OFB 2020). However, this ratio approach is considered to be insufficient to take into account biodiversity issues because it is based on a very incomplete description of the biodiversity involved, and it does not verify the equivalence between losses and gains (*Ibid.*). The reasons given to justify this gap between the recommendations and practices are a lack of human and financial resources, a lack of time in the naturalist consultancies to carry out in-depth studies and, lastly, a lack of regulatory ambitions.

However, in France, as in many other countries, the application of the mitigation hierarchy is governed by numerous regulatory principles:

– a strict *ecological and biophysical equivalence* (in-kind, as mentioned in the section above);

– an *ecological and regulatory additionality* provided by the offsetting measures to achieve the NNL objective of biodiversity;

– *the avoidance of temporary losses* between the provision of ecological gains and losses related to the impacts of the development project on biodiversity;

– a *strong geographical proximity* between the losses and the gains of biodiversity provided by offsetting measures;

– the *effectiveness* of offsetting measures, which implies the implementation of *monitoring and control* measures for the gains generated throughout the duration of the developer's regulatory commitments[10];

– the consideration of *risks of failure* in sizing offsetting measures associated with ecological uncertainties (which should result in an increase in offsetting needs, compared to the gross calculation made);

– a *technical, financial and organizational feasibility* of the ecological actions carried out in the offsetting.

10 These requirements are included in the prefectural decree that authorizes the issuance of the project's building permit, which can then serve as an enforceable document if the commitments are not met.

Theoretically, compliance with these various regulatory requirements should ensure that ecological issues at territorial scales are truly taken into account to achieve the objective of NNL, but the practice is different, as we have just pointed out.

In this respect, the academic literature is unanimous on the significant gap that exists between regulatory expectations and requirements, and the realities of the application of the NNL policies at the international scale (Maron *et al.* 2015). Indeed, in practice, the concrete implementation of the mitigation hierarchy does not seem to achieve the stated objective of NNL; the literature points out that, on the contrary, it results, in many cases, in a *net loss of biodiversity* (Calvet *et al.* 2015c; Maron *et al.* 2015; Moreno-Mateos *et al.* 2015; Weissgerber *et al.* 2019; Simmonds *et al.* 2020; Sonter *et al.* 2020). This is partly due to a lack of integration of ecological knowledge (Devictor 2018; Chassé *et al.* 2020).

Faced with these gaps between theory and practice, it seems legitimate to ask what place scientific ecology occupies in the application of this tool, which is supposed to meet the challenges of biodiversity conservation. In the remainder of this chapter, I propose to draw on the literature of scientific ecology in order to analyze how the latter clashes with the practice and the very principle of biodiversity offsetting around the *baseline* condition. In theory, this is a sine qua non condition for the practice of NNL, since impacts, losses and gains should only be evaluated in relation to baselines of the ecosystems involved (Bull *et al.* 2014; Maron *et al.* 2015). However, this baseline condition can involve different approaches to biodiversity: a stable or a declining biodiversity trajectory, a more or less anthropized or natural ecosystem, a past, present or future state of biodiversity, etc. In any case, the choice of the baseline is particularly decisive in the ecological actions that will be carried out subsequently, and therefore in the way in which the biodiversity offsetting policy will be conducted. Thus, beyond the ecological issues, the social and ethical considerations surrounding these choices and their consequences on the practice of biodiversity conservation will also lead me to call upon the human and social sciences in the second part of my work.

8.3. The question of a baseline in NNL policies: between ecological and socio-economic perspectives

In order to show the topicality and importance of the issues surrounding the baseline notion in the application of public conservation policies, here is a translated extract from Senator Dantec's report on the reality of the application of biodiversity offset measures (Dantec 2017, Volume 1, p. 74):

> Three points should be the subject of particular attention [...] including the definition of a *baseline scenario*, corresponding to the absence of a project and allowing, by way of difference, the analysis of the impacts of the latter on the environment and particularly on biodiversity. On this last point – the baseline scenario – the above-mentioned[11] directive states in its considerations that, among the various reasonable alternatives, there should be, where appropriate, an overview of the likely evolution of the current state of the environment in the absence of implementation of the project (the baseline scenario), in order to improve the quality of the environmental impact assessment procedure and to allow for the integration of environmental aspects at an early stage of project design.

8.3.1. *Ecological perspectives*

The baseline notion occupies a central place in the application of the mitigation hierarchy at several levels. Firstly, in the assessment of the *significance* of the impacts of development projects, which, if applicable, will "trigger" the rest of the procedure and the application of the mitigation hierarchy. If protected species or habitats are found, the baseline will also come into play in the assessment of the *possibility of derogating from the regulatory prohibitions* on the destruction of protected species (which ultimately leads to offsetting obligations, if applicable). Finally, the baseline will have to be mobilized in the context of *sizing* the ecological equivalence in order to aim at the NNL objective, as mentioned in the previous section.

Let us begin by returning to the question of assessing the *significance* of impacts. In the case of the environmental assessment of development projects, the mitigation hierarchy applies to impacts on biodiversity that are qualified as "significant" or "notable". An impact is qualified as "significant" when it undermines the biodiversity conservation issues it affects, or its role in the ecosystem then deemed a priority either locally or on a national scale (Biotope and DREAL LR 2016; Bezombes 2017). In this case, developers are obliged to apply the mitigation hierarchy, and after avoidance and reduction of these significant impacts, those that remain are then qualified as *significant residual impacts* and must be offset.

11 Directive 2014/52/EU of the European Parliament and of the Council of April 16, 2014 amending Directive 2011/92/EU on the assessment of the effects of certain public and private projects on the environment.

Concerning the protection of protected species, since the regulatory order of 2007[12], the law provides for the possibility of derogating from the prohibition on the destruction of species protected under European directives if "the derogation does not harm the maintenance of the species or the natural habitat in a *favorable* conservation status". In this case, it is possible to destroy the protected habitats and species in question, but it is still necessary to offer *offsetting*.

The conservation status of habitats or species is defined by Article 1 of the "habitats, fauna and flora" directive (92/43/EEC) as follows:

> the sum of all influences acting on a natural habitat and its typical species that may affect its long-term natural distribution, structure and functions as well as the long-term survival of its typical species within the territory referred to in Article 2.

It is also useful and interesting to specify that this possibility of derogating from the destruction of protected species was introduced in the law in 2007 in order to be able to officially authorize the killing of wolves, which are protected species. Since 2007 and this authorization of destruction of protected species, which implies the realization of offsetting, the number of files of requests for derogation has soared. It went from 44 files in 2006 to 197 in 2011, *i.e.*, four and a half times in five years (Regnery *et al.* 2013). This result reveals, in particular, a failure in the application of the mitigation hierarchy in France, which focuses only on protected species (*Ibid.*; Guillet and Semal 2018).

Finally, the regulatory principles that govern biodiversity offsetting aim to ensure the *additionality* of offset measures, in order to counterbalance the ecological losses related to the projects' impacts (in a perspective of achieving NNL). This objective of NNL implies that it is necessary to generate, through offsetting measures, biodiversity gains on a site, in order to increase its ecological "level" compared to a *baseline* (Maron *et al.* 2018). Meeting this condition then implies choosing baseline states at two levels: firstly, to assess the effects of development project impacts on the conservation status of species or habitats (losses), and secondly, to then calculate ecological gains provided by offsetting actions. The choice of these baselines is therefore crucial in the practice of the mitigation hierarchy. Indeed, depending on the approach chosen to define this baseline, "ecology" will not be called upon in the same way. For example, the baseline may take into account an ecosystem trajectory, and approaches of varying complexity to

---

12 Order of February 19, 2007 which sets the conditions for requesting and examining exemptions defined in 4° of article L. 411-2 of the French environmental code concerning protected species of wild fauna and flora.

the dynamics and functionality of ecosystems and their species (life cycle, potential biodiversity, etc.).

In practice, in order to evaluate the impacts of a development project, the environmental assessment begins with the realization of an *initial state* on the site targeted by the project. The purpose of the analysis of the initial state is to define the *baseline* of the site before development, in order to be able to identify losses, but what is the time scale to choose? The baseline could, for example, be set at the state of biodiversity present decades before, in order to truly reflect the reality of the trajectories and dynamics of the ecosystem, and to take into account the repeated impacts of the phenomena of anthropization in the area concerned. In current practice, the baseline considered is that *present* at the time of the environmental assessment of the project, *i.e.*, at the time when the development permits are requested. The geographical location of the project is generally already pre-established and only needs to be validated following the impact assessment procedure and the application of the mitigation hierarchy measures. This practice reflects a very static and localized view of biodiversity around the project in question. In this case, any biodiversity lost (*e.g.*, certain species in a forest) must be replaced in order to maintain the quantity, the level of biodiversity at its *current* level, *i.e.*, at the time of the initial status assessment. Thus, the biodiversity that constitutes the baseline of the site is the one present at the time of the field inventories carried out by the naturalist consultancy. Levels of conservation issues are then added to these field inventory results in order to determine the levels of impact according to the methods presented above (methods 2 and 3). The naturalist consultancy relies on reference documents in order to establish these levels of issues/impacts, mainly based on the lists and reference documents that classify species according to their level of extinction at the national or local levels[13].

Some ecologists have looked more closely at this issue of the choice of baselines in the context of NNL policies. The work of Martine Maron and Joe Bull has made a significant contribution to this literature, notably by studying the notions of "counterfactual scenarios" in the offset policies implemented in Australia. These scenarios consider that the baseline of biodiversity is that which would exist if the project and its offsetting did not take place by anticipating, through temporal projections, future trajectories of biodiversity. In their analysis of the Australian case, these authors have shown that taking these counterfactuals into account could ultimately work "against" biodiversity. Indeed, when an assumed trajectory of biodiversity decline is taken into account, a net gain from offsetting may correspond to maintaining the current state of biodiversity (Maron *et al.* 2015). If the rate of

13 For example, the IUCN red lists, and national lists such as the "Prioritization of protected species" carried out by local regulators with the support of local scientific committees (*e.g.*, the DREAL and the CSRPN).

biodiversity decline considered in the baseline scenario is high, even insignificant ecological actions can generate large offset credits, thus legitimizing even greater biodiversity destruction. Thus, instead of achieving an NNL of biodiversity via offsetting policies, taking these counterfactuals as baselines results in net biodiversity losses (*Ibid.*).

The biodiversity offsetting practices are often based on the concept of *averted loss* and, despite its obvious weakness, is the most internationally developed approach to calculating biodiversity offsets (the other being ecological restoration actions) (Simmonds *et al.* 2020). Indeed, a global state of the art, based on more than 12,000 offset projects, found that about 66% of them were based on this avoided loss approach, either exclusively or in combination with other approaches (Bull and Strange 2018). The rationale is that the protection provided by the offset actions achieves a benefit by avoiding a loss or decline in biodiversity that would otherwise have occurred. This practice of offsets is referred to as an "avoided loss" (also known as a protection offset) (*Ibid.*). Many offsets are sized according to this type of approach, which aims to propose only preservation actions as offsets (Simmonds *et al.* 2020). The latter are strongly supported internationally by financial institutions[14], non-governmental organizations (NGOs), such as the Business and Biodiversity Offsets Programme (BBOP)[15] and the International Union for Conservation of Nature (IUCN) and countries that play a key role in shaping NNL policies (Hrabanski 2015; Maron *et al.* 2018).

This approach is also used in other biodiversity conservation tools, the most emblematic of which are the payments for ecosystem services (PES), which are often mistakenly equated with biodiversity offsetting (Vaissière *et al.* 2020). In the PES system, landowners are paid not to destroy the natural environments they own, based on the assumption that they would otherwise destroy them.

In other works such as Bull *et al.* (2014), the authors emphasize the importance of the choice of baselines to evaluate the *performance* of conservation actions in the context of NNL policy. In this work, the authors show the need to define "reference frameworks" that can be either baselines, *i.e.*, a state of biodiversity assessed at a given point in time, or *counterfactuals*, which are based on scenarios without human

14 For example, the International Finance Corporation (IFC) Performance Standard 6 refers to no net loss through the implementation of offsets through the "avoided loss" approach in natural habitats (IFC 2012).

15 BBOP is an international NGO created in 2004 under the impulse of two NGOs, Forest Trends and the Wildlife Conservation Society. BBOP brings together nearly 80 actors from the political, governmental, private, financial, scientific and conservation spheres, with the objective of disseminating and promoting principles and good practices related to NNL policies.

modification (impact and offset). In this work, the authors developed models to study the implications of establishing these different baselines in regions that are subject to various biodiversity trends and anthropogenic impacts, as illustrated in Figure 8.3. The results showed that achieving the NNL target was highly dependent on the interaction between the baseline and background biodiversity trends. With a baseline approach, offsets were less likely to achieve NNL where biodiversity was already declining, in contrast to areas where biodiversity was stable or increasing. However, with a no-development counterfactual, NNL was only achievable where biodiversity was declining. Otherwise, they show that preventing development was the best scenario for biodiversity. When only development and offset areas were considered, offsets were sometimes able to achieve NNL, but not at the territorial scale (only on the area considered). These authors conclude on the importance of the choice of the reference framework that determines the feasibility and the effort required to achieve the objectives when designing and evaluating biodiversity offset programs. They emphasize that this choice is therefore of fundamental importance for the definition of conservation policies, because their results highlight situations in which biodiversity offsets may be an inappropriate policy instrument.

The criticisms of these counterfactual approaches, which are based on scenarios of biodiversity decline, are numerous. Firstly, the results are relative to a dynamic counterfactual trajectory that cannot be known in advance, only estimated. Secondly, biodiversity decline may continue even if a project results in NNL relative to a declining counterfactual. Furthermore, the construction of robust counterfactuals is conceptually complex and can be very data-intensive, leading to simplification choices, depending on the data available. Thus, the type and amount of offsets required are very sensitive to assumptions about the counterfactual path. It is then relatively easy to manipulate the counterfactual and its outcome. Finally, the end point of the biodiversity trajectory is unknown (Simmonds *et al.* 2020).

Simmonds *et al.* (2020) continued these reflections by proposing to differentiate conservation issues according to the spatial scales considered. They propose to work, on the one hand, with the *jurisdictional* scale, *i.e.*, that of the state or the regions that set the frameworks, rules and stakes for the protection of species, and, on the other hand, with the scale of the *development project* for which the offsetting is designed. These authors show that only a cross-cutting approach that makes it possible to integrate national, or even transnational, stakes into the sizing of offset actions at the project level can truly achieve relevant biodiversity conservation. In this perspective, they have developed a conceptual framework that aligns the objectives of project-level offset actions with the conservation objectives at the jurisdictional level. They then propose different ways of approaching the baseline, depending on the objectives envisaged, by proposing two possible paths: maintaining or improving the existing situation (Figure 8.3).

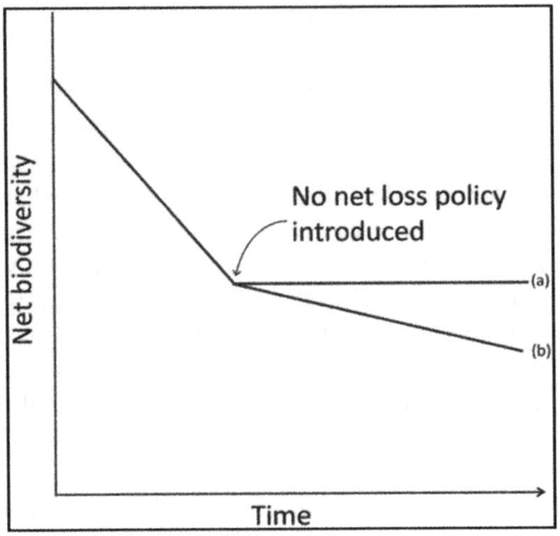

Figure 8.3. *Conceptual approach to the baseline notion in no net loss policies (source: Simmonds* et al. *2020)*

EXPLANATION OF FIGURE 8.3.– The term "no net loss" refers to different baseline scenarios. At the jurisdictional level, the term "no net loss" implies that the loss is halted in absolute terms relative to a fixed baseline – *i.e.*, all biodiversity losses are offset by gains of the same magnitude, thus maintaining biodiversity at the same level as before the loss (a). However, in reality, no net loss commitments often require that individual projects result in no net loss relative to a declining baseline, protecting biodiversity that might otherwise be lost in the future due to unregulated impacts ("avoided loss") (b). Such NNL at the project level results in continued biodiversity loss at the jurisdictional level, albeit at a slower rate.

In general terms, in all cases of the application of NNL policies, the baselines considered are either present or future, but never *past*. In this sense, I find it relevant to invoke the notion of "shifting baseline syndrome" proposed by Daniel Pauly (1995)[16]. With this notion, Pauly underlines the fact that we adjust our baseline level to the *present* states of biodiversity conservation, without taking into account past states. In a continuous regime of biodiversity destruction, each new generation of humans adopts a lower value as a biodiversity baseline than before. The baseline level then mechanically "shifts" towards less and less biodiversity considered. Applied to the case study of NNL policies, this means that each new generation of humans has to rely on less and less biodiversity, and thus offsets, to achieve its goal. Indeed, by focusing on the biodiversity states present in the assessment of ecological

16 On this notion, see also Chapter 4, by C. Moreau *et al.*

impacts and gains without taking into account the trajectories of *past* ecosystems, the baseline level chosen is necessarily based on increasingly low levels of biodiversity. However, an awareness of the real impacts of recent decades of development on biodiversity would allow us to avoid this effect and syndrome. If we push this logic and really consider past baselines in the environmental assessment of development projects on ecosystems by moving the cursor, not at the time of the initial state, but decades earlier, we would be faced with ecological losses so significant that no offsetting would be feasible or would be, with costs and investments, too significant. Moreover, this also raises questions of political and social responsibility with regard to these issues, which obviously go beyond the project developer alone. However, beyond knowing who should bear this burden, such an evaluation logic would allow for greater collective awareness of the ecological impacts of our contemporary social models.

The concept of *cumulative impacts* attempts to provide answers to these problems of taking into account multiple projects that follow one another in the same territory. These approaches, which are still relatively poorly developed in the current biodiversity offset practice, involve "zooming out" from the project's footprint to look at all the impacts generated by the development in *space* and, if possible, in *time*, on a territorial scale. The underlying idea is that, taken separately, each impact may have little effect on biodiversity (on the dynamics of a species population, for example), and thus generate a low need for offsetting, whereas taken as a whole, the accumulation of impacts may ultimately have serious consequences on biodiversity. The literature mentions "death by a thousand cuts" situations to illustrate this (Whitehead 2017). Notions of "thresholds" can be useful here to approach this problem. For example, the thresholds of minimum surface areas necessary at the scale of a territory to ensure the lifecycle of a species.

In these approaches, no account is taken of *past states* of biodiversity, the baseline remaining the *present* state. However, when it comes to ecosystems that have taken hundreds of years to form (such as peat bogs, for example), quantifying the impact on these ecosystems from the present moment omits any ecology of these ecosystems. This is all the more true when it comes to regulatory commitments to maintain offsets for an average of 5–30 years (which is then considered "long term") (Guillet and Semal 2018).

Another criticism of the practice of offsetting is the almost exclusive focus on *protected biodiversity* and the *local* issues surrounding this biodiversity (Regnery *et al.* 2013; Simmonds *et al.* 2020). However, from a regulatory point of view, and in accordance with the main international principles of application of NNL policies, biodiversity offsetting should concern impacts on *all components* of biodiversity:

> in terms of species diversity, functional diversity and composition, ecological integrity or condition, landscape context (*e.g.*, connectivity, landscape position, adjacent land uses or condition, patch size, etc.), and ecosystem services (including people's use and cultural values) (BBOP 2012).

Protected species may show positive population dynamics locally but be in a very critical conservation status at the national or even international level. This is, for example, the case of the Little Bustard (*Tetrax tetrax),* which has a very positive population dynamic in the Mediterranean regions of southern France (former Languedoc-Roussillon and Provence-Alpes-Côte d'Azur regions) (Devoucoux *et al.* 2019). Indeed, with the very "protected species"-centered approach in the mitigation hierarchy practice in France, the Little Bustard represents the species that elicits the most offset measures in these regions, being very widespread locally, and therefore the object of many impacts. However, although this species has a positive population dynamic in these regions, it is classified on the IUCN red list at the international level and as an "endangered species" at the national level in France. Another consequence of this focus on protected species is that a significant portion of biodiversity, which is not listed or targeted by a specific protection policy, is neither considered in the impact studies nor targeted in the offsetting. However, these species, often referred to as "common" or "ordinary" and considered abundant, play an essential role in maintaining ecosystems and ecological processes, and could ultimately become threatened if they are not considered (Calvet *et al.* 2015c; Devictor 2018).

Faced with the practices of the mitigation hierarchy, other questions related to ecology seem legitimate to ask, such as taking into account the spatio-temporal dynamics of the ecosystems concerned, both on the impacted sites and on those that will be used for future offsetting. Indeed, while an initial assessment is almost always carried out on the impacted site, this is not often the case on the site to be used for offsetting. Few studies mention an environmental assessment of the offset sites, especially when it is a question of sizing the equivalence on the basis of surface ratios, an approach that is still widely used, as mentioned above. In the case where initial states are carried out on offset sites, in all cases, only the present and visible states of biodiversity are considered.

In order to compensate for this lack of "*dynamic*" consideration of biodiversity, more and more scientific work is developing modeling approaches in order to simulate the future states of the territories considered, and to anticipate the effects of the predicted impacts on biodiversity (Bigard *et al.* 2017; Calvet *et al.* 2019a; Tarabon *et al.* 2020). On the other hand, work also highlights the importance of considering *potential* biodiversity with respect to the habitats considered (Regnery *et al.* 2013). Indeed, any observed ecosystem is the seat of past and future dynamics of biodiversity, and also of a non-visible biodiversity (known as dark diversity) that strongly participates in ecological processes (Pärtel *et al.* 2011). However, in practice, the constraints in terms of human, financial, technical and time resources in the naturalist consultancies result in the development of compromises between taking into account global approaches to biodiversity issues and "simple" observation data from very localized fields around the project impact area (Dauguet 2015).

Faced with these issues, analysis of the application of NNL policies and biodiversity offsetting may lead to the conclusion that there is a form of incompatibility between the theoretical assumptions required for the practice of offsetting (the substitutability of ecological entities by others) and the knowledge developed over decades in scientific ecology, that teaches us that each ecosystem is unique and irreplaceable (for more details, see Calvet *et al.* (2015c) and Devictor (2018)). From this perspective, from an ecological point of view, the practice of offsetting fundamentally appears to be "a-ecological", as it leaves out or oversimplifies the biophysical properties and processes that it is nevertheless supposed to compensate for, to achieve "no net loss".

Thus, these analyses lead to fundamental questions that go beyond the framework of scientific ecology, such as: can we replace one ecosystem with another? What biodiversity trade-offs are socially acceptable? Ultimately, the answer to these questions also requires the social sciences to be called upon.

8.3.2. *Socio-economic perspectives*

In economics, the question of the reference framework applied to the NNL policy involves several research areas. Initial works have posed this question of the choice of baselines with regard to the objectives that the policy may aim to achieve, asking "No net loss of what?" (Calvet *et al.* 2015c). These works highlight that depending on the policy objectives targeted in NNL, the kind of equivalence, offsets and metrics to assess them will strongly differ. For example, an NNL objective based on maintaining human well-being may be translated in an equivalence in terms of *utility*. This will, in turn, lead to the design of biodiversity offsets in terms of capital (natural, manufactured) which can be evaluated on the basis of cost–benefit analyses, for example. Biodiversity losses would then be seen as costs that could be offset by benefits from future developments. On the other hand, a policy based on an NNL of species and habitats would imply the search for a *biophysical* equivalence requiring offsetting of the same ecological entities lost via the mobilization of ecological indicators. These different approaches reflect disparities in the way of considering "the environment" within economics frameworks, which have led to the distinction of two currents: environmental economics and ecological economics[17].

Between these two approaches to NNL, these works lead us to consider two other categories that tend to develop increasingly in the practice of offsetting: *functional* equivalence and equivalence in *ecosystem services*. Functional equivalence leads to a focus on ecosystem functions, while ecosystem service equivalence focuses on sizing the losses and gains through the prism of the services they provide to humans. Many studies warn of the risks of such approaches for biodiversity conservation, which would lead, for example, to the replacement of complex ecosystems (such as wetlands) by other, more "simple", ecosystems that provide the same functions (*e.g.*, purification) or the same services (*e.g.*, flood regulation). While such approaches tend to develop strongly in the offsetting practice, the conclusions of these works lead to consider them in a complementary manner with other biophysical approaches that remain focused on species and habitats as such (Calvet *et al.* 2015c; Jacob *et al.* 2016).

Indeed, the work in scientific ecology mentioned in the previous section shows the difficulties, or even impossibilities, of replacing ecological entities by others, even if it is a *like-for-like* and *in-kind* equivalence. Therefore, considering functional equivalence, equivalence in ecosystem services or in utilities would necessarily result in a net loss of biodiversity (*Ibid.*).

Other works of economists have focused more on the notion of a baseline based on individuals' *perceptions*. Gregory *et al.* (1993) introduced the idea of a baseline by studying the effect of past environmental "states" on individuals' willingness to support mitigation policies. In their work, they conducted an experimental analysis in which students were asked to assess the desirability of environmental improvements in different areas. These environmental improvements were presented either as improvements to the current situation or as restorations conducted after degradation of the state of the ecosystems (without specifying responsibilities). The results of this work reveal that respondents preferred to support a return to the original state following a previous environmental loss rather than a net environmental improvement of a current situation. Gregory *et al.* (1993) refer to prospect theory and loss aversion to explain respondents' greater willingness to support a "return to normal" policy rather than an "improvement" policy. Restoring the environment to the way it was in the past appears to be more satisfying for "loss

17 For more on this topic, see Calvet, C. (2015) and Boisvert, V. (2017). Économie de l'environnement ou économie écologique ? In *Humanités environnementales. Enquêtes et contre-enquêtes*. Blanc, G., Demeulenaere, E. and Feuerhahn, W. (eds). Publications de la Sorbonne, Paris, Chapter 9.

averse individuals" than considering gains for a current situation. Here again, the question of the choice of the "past state" as the baseline scenario arises.

In environmental economics, a theoretical presupposition is always mobilized when it comes to the economic valuation of biodiversity: the discount rate. This rate is supposed to allow for the inclusion of time in the valuation of biodiversity (or its changes) in economic calculations (Chevassus-au-Louis *et al.* 2009). This economic reasoning, derived from thinking about financial flows, seeks to account for "the preference that individuals have for the present" (*Ibid.*). In other words, the discount rate is the rate of substitution between the present and future consumption of resources and services. This rate states that the loss in a given year is passed on to the following year. It is commonly set at 4% in the short to medium term (decreasing in the long term) in the case of environmental valuation in calculations of ecological losses and gains. This factor, which, in principle, provides an incentive to reach the end of losses more quickly, tends to ignore the long-term consequences of ecological losses.

In the face of the theoretical impasses and ecological risks highlighted in this chapter, significant controversies have emerged that also raise ethical limits to NNL policies. Without drawing up an exhaustive list, I will present some of them in the remainder of this chapter in line with the developments I have proposed in the preceding sections. The identification of all these limits will then lead me to ask the question of *why* in the conclusion of this chapter. *Why* develop these practices in spite of all the scientific controversies that highlight failures in achieving biodiversity conservation objectives? Answering this question requires going to the past in order to understand the origins of these policies and the objectives and values that motivated them in the past and still today. We will discuss these elements around socio-political perspectives.

8.4. Implications of NNL policies for biodiversity conservation: ethical and political perspectives

The ethical issues raised by NNL policies are diverse, and I propose to address some of them in this section. Firstly, the processes of equating biodiversity by the calculation methods described above, which make it possible to exchange biodiversity A for biodiversity B, are particularly criticized. Mainly mentioned in these criticisms is a form of "commodification" of nature that renders it abstract and makes it lose its materiality and biophysical reality (Dauguet 2015; Devictor 2018). Some authors point out that this "commodified" biodiversity in units opens the door to a selling of nature, even if it does not necessarily involve the production of market goods in the theoretical sense (Castree 2003; Maris 2014; Boisvert 2015; Dauguet 2015).

Many authors have also highlighted a detour of purposes in the practice of offsetting that results in the loss of the initial objectives, namely the conservation of biodiversity. They underlined the exacerbated part given to offsetting to the detriment of the other stages of the mitigation sequence, both in practice and in the human and financial resources available to the state services (Guillet and Sémal 2018). The authors also show that offsetting precisely diverts conservation issues towards a development logic (Gordon *et al.* 2015; Apostolopoulou and Adams 2017). On the basis of this work, the authors go so far as to question the validity of this tool for conservation, even seeing it as a way to "pulverize" biodiversity by agreeing to destroy it without actually compensating or protecting it; they talk about "bulldozing biodiversity" (Spash 2015).

Offsetting is a particularly relevant object of study because it crystallizes scientific and societal debates on the future of biodiversity and on ways to ensure its protection. In this sense, it seems useful to link the development of this policy instrument to a new form of biodiversity conservation, which has been emerging for many years in political arenas with New Public Management (Bouleau and Pont 2015) and that resonates within the conservationist community with the New Conservation Movement (see Doak *et al.* 2013; Calvet *et al.* 2015a). This work highlights the managerial logic behind the development and promotion of these tools, which simply entrench a certain way of looking at human and non-human relations, in a form of domination of the former over the latter. The work carried out in political ecology and in the history of science makes it possible to trace the origins of NNL policies in order to better understand *why*. They highlight an ideological legacy that seems to be perpetuated in the biodiversity conservation practices of contemporary Western societies (Hough and Robertson 2009; Bonneuil 2015; Calvet *et al.* 2015a). Indeed, this work reveals that NNL policies were originally thought of by the U.S. government as flexibility policies, developed with the goal of countering the environmental movements that were gaining momentum in the 1980s following early ecological awareness. Historians call this movement the environmental backlash (Bonneuil 2015). Offsetting made it possible to show that environmental issues were taken into account while allowing developers to continue to carry out their development projects (*Ibid.*).

The analysis of the geopolitical context surrounding the emergence of this policy allows for a better understanding of why and how the choices to apply the mitigation hierarchy at the international level are made. This detour, or rather return, to the origins of this sequence underlines the importance of adopting a reflexive approach when studying, or using, objects that are far from neutral, but normatively charged, and whose consequences for biodiversity can be significant.

8.5. Conclusion

In conclusion, NNL policies are gradually becoming a part of biodiversity conservation practice on an international scale. However, their use raises important theoretical, practical and ethical questions.

Theoretically, the presupposition on which they are based, *i.e.*, the ecological equivalence between a destroyed biodiversity and a biodiversity restored in the framework of offsets, invites us to remobilize knowledge from scientific ecology in order to discuss their validity. The work carried out on this subject underlines the sources of strong incompatibilities between the objectives targeted – the absence of a net loss of biodiversity – and the mobilization of ecological knowledge, such as taking into account the integrity of ecosystems and the complexity of life.

In practical terms, the operational implementation of the evaluation of ecological equivalence leads to a high degree of heterogeneity in practices, both within and between countries. Unfortunately, the resulting calculation methods are not always ecologically satisfactory, in terms of taking into account the temporal and spatial dimensions of biodiversity. Paradoxically, despite the recent regulatory reinforcement of biodiversity offsetting, its application is still the subject of compromises and agreements between stakeholders, often to the detriment of biodiversity.

These results reinforce the criticisms of offsetting policies, which necessarily lead to ethical considerations, particularly with regard to the choices to be made in terms of biodiversity conservation actions and land-use strategies.

A historical detour around the emergence of NNL policies in the United States reflects this persistent tension between economic development and biodiversity conservation, which offsetting aims to overcome and which it nevertheless only crystallizes.

In this sense, NNL policies offer the opportunity for citizens to re-engage with these issues of choice between economic development and biodiversity conservation with the ambition of re-politicizing them.

I will conclude by saying that, beyond being nostalgic for the *past*, it is sometimes necessary to go back to it to better understand the *present* and to arm ourselves to defend the protection of nature in the *future*.

PART 3

Examples of the Use of Baselines

9

The Variability of Baselines Mobilized in Littoral Protected Areas: The Anthropocene as a Dividing Line?

9.1. Introduction

Over the past 40 years, the protection of natural areas on the northwestern European coast has greatly increased, reflecting both a strong diversification of protection regimes and an increase in the number of protected areas. From one country to another, nature reserves, national parks, land and/or landscape protection, Littoral Protected Areas and Natura 2000 sites now form a complex network from a regulatory and organizational point of view. Paradoxically, while the mobilization of baselines is at the very foundation of all these conservation projects, however heterogeneous they may be, of their heritage, scientific and socio-political legitimacy (Moreau *et al.* 2019), it remains rarely explained and defined (Bull *et al.* 2014), let alone argued. The notion is in fact extremely plastic, covering very different realities from one site to another. Thus, the baseline can constitute the desired horizon and objective of a conservation project, just as it can be mobilized as a tool to measure the dynamics of protected ecosystems, as well as to evaluate the effectiveness of the management practices deployed. Given the complexity of ecosystem monitoring, the choice of a baseline and its uses raise a series of questions that have long been widely debated in conservation ecology, as well as in the community of protected area managers: what objective criteria can be mobilized to choose a baseline among a wide range of possibilities? First, since ecosystems are dynamics in nature, any conservation project defined in terms of a specific baseline may tend to freeze the ecosystem in space and time, whether it is a baseline to be reached or a baseline from which a manager does not wish to deviate too much. Second, how can we distinguish the effect of management measures in a protected

Chapter written by Vincent ANDREU-BOUSSUT and Céline CHADENAS.

area from the dynamics that are specific to ecosystems, or dependent on changes in ecological conditions imposed outside of the protected area? Finally, in many cases, the baseline often coincides with the initial state of the protected area, which justifies its protection, even though this initial state may not, in many cases, be the optimum state of conservation of these ecosystems or express their full ecological potential. The choice of a baseline in conservation projects is therefore not self-evident and ultimately constitutes a value judgment (Bull *et al.* 2014). It is a scientific construct, depending on the knowledge of ecosystems, their dynamics and measures of their conservation status, just as it is a socio-cultural construct, depending both on the territorial context in which the protected area is inserted, as well as on the manager, their culture, their scientific background and the organization they work for or their links with territorial stakeholders.

The protected areas along the northwestern European coastline, and especially the nature reserves, thus constitute a favored scene for observing the way in which baselines are mobilized and constructed, and the questioning that is imposed, in particular in the short and medium terms, by sea level rise. The choices of baselines from which conservation projects are deployed are therefore very diverse, but can be ordered according to the time horizon mobilized. While the most common position consists of "matching" the contemporary arrangement of species and habitats whose value justifies the creation of the protected area, certain choices may diverge either in the direction of a desired return to a previous historical state (supposed to be a better state of conservation), or in the direction of an adaptation or accompaniment of the evolution of the ecosystems in a future horizon. We propose here a synthesis of five types of baselines, as they can be mobilized in coastal protected areas (see Figure 9.1), modifying and adapting the typology proposed by McNellie *et al.* (2020). This categorization distinguishes between prehistoric ecosystems, historical ecosystems that predate the intensification of development, contemporary ecosystems as nature–culture hybrids, ecosystems whose evolutionary dynamics are controlled (particularly in the context of adaptation to sea level rise) and novel, freely evolving ecosystems. Each baseline is linked to a precise conceptual framework and is based on specific conservation objectives, implying differentiated management methods whose degree of intervention varies widely from one approach to another. These baselines are distinguished, for example, by the conservation approach that underlies them, either a compositionalist approach that favors the structure of ecosystems (levels of biodiversity, target species, etc.) and implies a more or less strong control of ecological dynamics, or a functionalist approach that targets the maintenance and/or deployment of ecological processes and thus the evolution of ecosystems (Callicott *et al.* 1999; Jepson 2016).

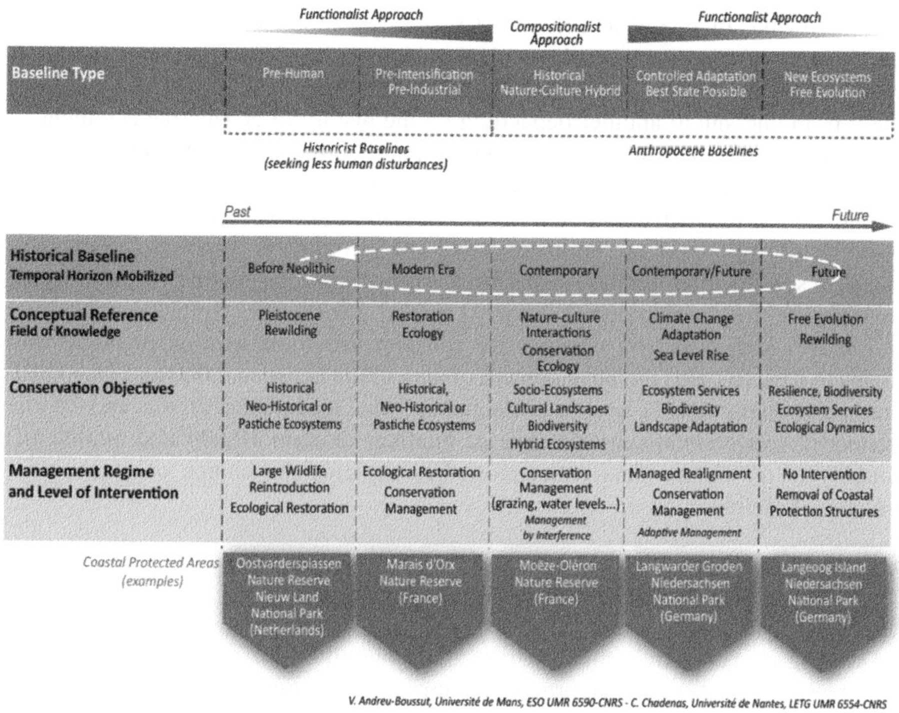

Figure 9.1. *Types of baselines used in coastal protected areas*

In this typology of mobilized baselines, the Anthropocene constitutes a dividing line revealing divergences in positioning. Thus, the choice of a historical baseline is generally guided by the search for less human disturbance, in a pre-Anthropocene configuration. Conversely, taking note of the role of disturbances as drivers of high levels of biodiversity or the fact that ecosystems are subject to global anthropogenic changes responsible for now irreversible evolutions requires a baseline to be defined as anchored in the Anthropocene, using the concept of Anthropocene baseline formulated by Kopf *et al.* (2015). While the choice of this baseline depends on the geographical, historical and local ecological context in which the conservation project is carried out, its positioning in relation to the Anthropocene also results from the understanding that protected area managers have of this new geological period, and their way of dating this tipping point in the functioning of ecosystems into a new regime dominated by human activities. Protected areas are ultimately not immune to the scientific community's debate over the triggers and dating of the origin of the Anthropocene (Lewis and Maslin 2015). For example, the most common and longstanding conception of the Anthropocene, which dates back to the beginning of the Industrial Revolution (Crutzen and Stoermer 2000)[1], involves the identification of historical baselines that predate industrialization or agricultural intensification. In reality, this historical baseline is not univocal since it can vary widely from one site to another depending on the history of coastal development. It all depends on whether the identified tipping point corresponds to the polderization and land reclamation of coastal marshes (mostly between the 17th and 19th centuries) or to the intensification of agriculture for moorland and coastal grassland environments, all the later in history that the geographic areas are peripheral. Finally, a second interpretation dates back to the beginning of the Anthropocene, much earlier in environmental history to the Pleistocene–Holocene transition (11,700 BP), depending on whether we consider the domestication of plants and animals to have initiated the development of agriculture between 9,000 and 10,000 BP (Smith and Zeder 2013) or the extinction of the megafauna at the end of the Pleistocene concomitant with the rise of hunting between 10,000 and 50,000 BP (Doughty *et al.* 2010). This conception thus explains the choice of much older prehistoric baselines.

9.2. The prehistoric baselines of paleo-rewilding

Drawing on paleoenvironmental data that now allow for greater knowledge of ancient environments, a new approach to nature conservation has favored prehistoric baselines for the past 30 years. Deployed under the name paleo-rewilding (Kopnina *et al.* 2019), this radical posture stems from the rewilding movements (Barraud *et al.* 2019). Such an approach implies the reintroduction of now-extinct fauna and their replacement by ecologically analogous species with the aim of reproducing environmental processes and reconstituting now-extinct ecosystems (see Chapter 5). Insofar as the most emblematic approaches generally target the reintroduction of species close to the megafauna of herbivorous mammals that disappeared at the end of the Pleistocene, this movement is also called Pleistocene rewilding (Barraud *et al.* 2019). Conjured up as a possible pathway to better conservation (Caro *et al.* 2011), its experimentation is nevertheless the subject of much controversy. In Littoral Protected Areas (LPAs), this prehistoric baseline remains very rarely mobilized, often because natural environments are considered too artificial to imagine a return to such an ancient time horizon, where the coastline occupies a position far from its

1 This conception of the Anthropocene is now complemented by the notion of the Capitalocene, which attributes significant responsibility for the emergence of this new geological era to the globalization of capitalism (Malm 2016). In light of the close intertwining of the global anthropization of ecosystems and racist, slaveholding and colonial histories over time, Haraway (2015) and Tsing (2017) also define the concept of the Plantationocene in reference to the implementation of the plantation regime as a triggering moment for the Anthropocene.

current position. Nevertheless, the most widely publicized Pleistocene rewilding experiment in Europe is coastal, even if its implementation was primarily a land opportunity.

Completed in 1968, the Oostvaardersplassen polder occupies 5,600 hectares reclaimed from the Markermeer to the northeast of Amsterdam. A nature reserve was created there in 1974 in view of the richness of the avifauna that had spontaneously established itself in the wetland (notably the Greylag Geese *Anser anser*). It is owned by the state, and its management is entrusted to the Staatsbosbeheer, a recently privatized public organization in charge of nature conservation in the Netherlands. The conservation project carried out there from the 1980s onwards owes much to the ecologist Frans Vera. This "life-size" experiment allowed him to test his theory of a post-glacial forest at the end of the Pleistocene that was less uniform and less closed than usually imagined, and whose landscape structure was shaped by large herbivores (Vera 2000, 2009; Barraud 2021). Species similar to the now extinct grazers were reintroduced and de-domesticated from 1983 onwards (Heck's Aurochs, Konik horses and Red Deer), and gradually allowed the reconstitution of a semi-open landscape. In this new "*Serengeti behind the dikes*" (Jepson and Ladle 2010), animal populations are also expected to achieve self-regulation, dictated by climatic conditions and available food supply, with dead animals being recycled by scavengers. Such a model is doubly fascinating because it illustrates the possibility of reconstructing pre-Anthropocene ecosystems and landscapes and recovering ecological processes free of human control (Balmford 2014). Lorimer (2015) describes this Oostvaardersplassen experience as a wild experiment that is itself wild because its heterodoxy lies in the way it has freed itself from the usual techno-administrative straitjacket of protected area management, being guided by surprise without a specific conservation goal.

While the Oostvaardersplassen Nature Reserve thus now serves as a benchmark for other nature[2] conservation projects, it remains extremely controversial for at least four reasons. The first reason is that the reconstructed historical and rewild ecosystems can be considered as "past imperfections" (Alagona *et al.* 2012) in that they now thrive in a very different Pleistocene climate, thanks to feral species, in an artificially controlled environment behind dikes and fences. According to this critique, the historical ecosystem can be considered as a pastiche. The second reason is the geographical context of its implementation, the scale of such a conservation project is presumed to be too spatially limited, in an overly artificial and developed territory. While the approach is inspired by early North American pioneering movements of rewilding fostered by the reconstitution of complete food chains

2 Oostvaardersplassen has largely inspired other *rewilding* experiments, notably at the Knepp Estate in Sussex, UK (Tree 2019).

allowing for the self-regulation of ecosystems (Soulé and Noss 1998), generally deployed under the 3 Cs banner of *core, corridor* and *carnivore* (Barraud *et al.* 2019), it neglects the integration of ecological connections at a larger scale, just as it replaces carnivores with herbivores as keystone species. The third reason is related to the high mortality caused by an overpopulation of herbivores in a constrained space with no opportunity for migration. This leads to a very strong local and national advocacy demanding that animals be given proper care, which explains ecologist Kopnina's argument for a fourth C, for *care,* and advocates for redefining a sensitive relationship with non-humans (Kopnina *et al.* 2019). The final reason is the absence of a management plan and precise monitoring of species and ecosystem dynamics for more than 20 years, whose regular evaluation and interpretation would have allowed for the refinement and guidance of different choices.

In the face of these criticisms, the interpretation and meaning of the Oostvardersplassen experiment undergoes a significant shift from a historicist reconstruction to one that is gradually described as an attempt to produce new ecosystems and arrangements of "wild" nature in the future opened up by the Anthropocene (Lorimer 2015; Clement 2021). More importantly, the conservation project has drastically changed its trajectory and the management of the reserve is now reoriented in two directions, one devoted to the conservation of wetlands favorable to waterfowl, while the second remains dominated by the grazing of large herbivores whose population is controlled by regular preventive culling measures (the meat is sold, by the way).

9.3. The historical baselines of prior states to development intensification

The reclamation of historic ecosystems is intertwined with the history of restoration ecology and ecological restoration experiments. In France, as in other countries along the northwestern European coast, experimental ecological restoration operations began in the late 1980s and early 1990s (Gallet *et al.* 2017) and expanded in the following decades, encouraged both by the improvement of knowledge, practices and the construction of networks of dedicated professionals (Clewell and Aronson 2010), and by the evolution of European public policies. In this respect, the European Habitats Directive of 1992, which initiated the construction of the Natura 2000 network, largely promotes ecological restoration with the objective of improving the conservation status of species or habitats with a European heritage at stake, just as the European Water Framework Directive of 2000 imposes the restoration of wetlands with the aim of regaining the overall quality of water (in particular through the principle of ecological compensation). For natural coastal environments, ecological restoration operations are becoming frequent, particularly for moors and cliff grasslands degraded by decades of trampling due to their frequentation (Enoul 1999; Gallet *et al.* 2011), and for wetlands for which the developments made must be subject to compensatory[3] restoration. Restoration ecology is thus at the origin of a new conservation approach, which can then promote the return to past states for ecosystems considered sufficiently damaged. In most cases, it is a question of restoring a prior state to the historical developments that were responsible for the "degradation" of the ecosystem, making it possible to regain, to varying degrees, more naturalness in the ecological processes, higher levels of biodiversity, heritage species in a better state of conservation or a landscape that is more socially desirable. In reality, restoration baselines are often divided between a more or less unknown, and sometimes idealized, historical state and current states (Donadieu 2002). The definition of a baseline is then guided by the historical and archeological knowledge that can be mobilized, but fundamentally remains an approximation of the desirable state, a standard chosen among several possible alternative states and accessible through a succession of steps called a trajectory (Le Floc'h and Aronson 1995). The choice among several possible horizons then depends on heritage issues and the financial and technical available means, but it is also ethically dependent on the local cultural and socio-political context (Higgs 1997).

In the Landes region, the Orx marsh is an exemplary case. The vast body of water and its associated marshes were partially drained as part of an ambitious agricultural development project during the 19th century and then put to the test of intensive agriculture, which finally collapsed in the early 1980s due to the cost of hydraulic maintenance that had become prohibitive. A large part of the agricultural domain was acquired by the Conservatoire du Littoral in 1989, and an ecological restoration operation was then proposed in the context of the creation of a national nature reserve in 1995. The ecological intervention was initially planned with the objective of "rejuvenating" the natural environments and helped to produce a new geography of the protected marsh. The reserve is compartmentalized into three sets of habitats and landscapes managed independently and which thus offer a testimony of the genesis and evolution of the Aquitaine fluvio-lacustrine systems (Baron-Yellès 2000): the central marsh, renamed the "original marsh", was restored with the objective of restoring the basin of lacustrine environments that existed in the Middle Ages, before the drainage and sanitation works, and which constitute many favorable habitats for Anatidae populations. In the northern part of the marsh, the wet meadow typical of the dykes built in the 19th century and of

3 In the 1990s, the construction of the Port 2000 container terminal of the Grand Port Maritime du Havre required the restoration of meanders and mudflats in the national nature reserve of the Seine estuary as part of ecological compensation. This was one of the first ecological restoration operations of coastal wetlands in France.

non-mechanized agriculture has been restored and renamed the "heritage marsh"; finally, the marsh-dam located in the south, renamed the "living marsh", has been entirely transformed to produce a natural palustrine environment favorable to small and large wading birds. A low water level is maintained in a largely reworked relief in order to promote the nesting of the avifauna. The Marais d'Orx bears witness to the mobilization of several historical referents, resulting in an "anachronistic situation juxtaposing landscapes that existed at different times" (Baron-Yellès 2000, p. 137). This reconstruction is obviously far from perfectly accurate from a historical point of view for at least three reasons. First, the 19th century grazed marsh was restored to when the marsh was dedicated to intensive agriculture, whereas it was historically more developed in the center of the marsh. Second, given the lack of historical knowledge about the functioning of the ecosystem before its agricultural development, another regional baseline was used as the objective of the restoration, the Barthes de Saubusse, located on the banks of the Adour. Finally, the "living marsh" was imagined by invoking non-regional baselines: deltas, large American marshes, Amazonian marshes", etc. Although historical baselines are sometimes mobilized to produce more desirable ecosystems, it is also sometimes a question of rediscovering agricultural practices that have disappeared and that are at the origin of singular ecosystems or favorable to certain species considered as heritage. In the United Kingdom, the Royal Society for the Protection of Birds (RSPB) has been implementing a policy since the beginning of the 1990s to improve the conservation of the corncrake *Crex crex*, whose populations were then in danger. The ecology of this species is largely dependent on open environments, but its conditions of survival have been largely damaged by the mechanization of agriculture since the 19th century. In its reserve on the Isle of Coll in the Hebrides archipelago in Scotland, the RSPB conducted experiments to measure the effects of non-mechanized agricultural practices on corncrake populations: season and rhythm of mowing, direction of mowing, etc. While this experiment revealed the most favorable agricultural practices for the species, it also restored pre-modern agricultural practices whose knowledge had largely been lost. The success of the RSPB's work then serves to support a national conservation policy of the corncrake deployed through the return of subsidized[4] historical agricultural practices.

9.4. The contemporary baselines as historical hybrids between nature and culture

Until recently, the majority of protected areas used contemporary baselines, defined more or less according to the initial state of the ecosystems and landscapes

4 Lorimer (2015) devotes a detailed analysis to the restoration and widespread use of conservation-friendly agricultural practices for corncrakes, which he describes as "securing pre-modern ecologies".

for which the evaluation of heritage values justified their protection. While such a baseline may be a certain *status quo*, it is nevertheless accompanied, if necessary, by the search for better states of conservation for certain target species or habitats. Thus, for natural coastal environments that are, to varying degrees, heavily influenced by the history of their agricultural uses, the baseline ecosystems are largely agro-ecosystems inherited from a traditional agrarian metabolism (Schnitzler *et al.* 2008a, 2008b). They are both cultural landscapes and biodiversity "factories", the specific assemblages they contain often being dependent on and induced by past or ongoing agricultural uses (see Chapter 11).

The conservation of such semi-natural ecosystems with high levels of biodiversity implies a quasi-permanent control of ecological processes, which are dynamics by nature. Inspired by a paradigm that assumes the existence of an ecosystem equilibrium (Blandin 2009), the principles of intervention are generally aimed at maintaining or blocking these systems at a desired stage, sometimes far from their potential climatic state because they are the result of specific disturbance regimes, even though these regimes can evolve eminently. This posture, often described as conservation management, is therefore a "management by interference" (Adams and Mulligan 2003), the degree of intervention of which varies from one site to another. Considered as eco-gardening (Lemoine 2005), interventionism can be intensive insofar as it is dictated by the heritage value of sub-climactic[5] ecosystems. The management of protected areas thus requires the maintenance or redeployment of agricultural practices considered to be at the origin of the production of these ecosystems (extensive grazing, mowing, etc.). Nevertheless, these interventions do not necessarily consist of recovering/simulating pre-modern agricultural practices, as their objective is to promote or improve a desired state of biodiversity and not to turn managers into farmers. Thus, an old salt marsh can be re-watered, not to produce salt, but to be favorable to avifauna. This hybrid management finally combines empirical knowledge of the functioning of natural environments, knowledge of past agricultural uses and ecological engineering knowledge. This management of biodiversity (Blandin 2009) promotes open environments (marshes, lawns, etc.), regular maintenance of water bodies to avoid their closure or the proper functioning of the hydraulic systems of marshes and polders, etc. Such control of ecosystems allows these protected areas to fulfil their regulatory role in the conservation of certain habitats and species, and also allows them to give strong social legitimacy to the conservation projects they carry out. Thus, the control of an emblematic biodiversity can be at the origin of tourism development, as in the case of the Marquenterre bird park in the Baie de Somme National Nature Reserve (Chadenas 2008; Chadenas *et al.* 2020).

---

5 Adams and Mulligan (2003) particularly document this intensive management and consider it dominant in protected areas in the United Kingdom.

This dominant "model" in the management of protected areas is increasingly being questioned on at least two interconnected levels. The first criticism concerns the initial state of protection as a baseline, insofar as this choice obscures past states or potential states that may be richer, and prevents the continuation of ecological dynamics that are in any case induced by natural processes and global changes. This debate is based on the notion of shifting baseline syndrome (Pauly 1995), as the contemporary baseline can be interpreted here as witnessing an ecological amnesia (Dubois 2012) that forgets the environmental history of a site and considers its initial state as highly desirable, while other more ecologically favorable states are forgotten (see Chapter 4). In this respect, the reference to open environments as factors of high biodiversity is regularly criticized (Génot 2010). However, this situation is not necessarily attributable to protected areas' managers, who would be affected by generational or personal amnesia (Papworth *et al.* 2009), but can be explained more by a lack of precise knowledge, for the majority of sites, on the past states of the ecosystems and by a logic of creation and operation of protected areas rooted in a heritage argument and in an accounting logic. The hierarchy of values that underlies the heritage regimes is also the subject of controversy because it leaves out ordinary nature and obscures the evolutionary capacities of ecosystems[6], enclosing the protected areas in a "museum bubble" with environments that do not evolve much. The second criticism concerns the interventionist nature of management methods, which can be considered as a desire to control a confined and domesticated nature, facilitated and enabled by technoscience (Schnitzler *et al.* 2008a, 2008b; Génot 2010). Such a critique is obviously linked to an ecocentric conception of a decolonized nature that is independent of its human uses, and to a questioning of the paradigm of ecosystem equilibrium (even if there may be states of equilibrium that are more or less long-lasting) in favor of a more evolutionary conception of ecosystem trajectories (Blandin 2009).

9.5. The negotiated, controlled and adapted baselines in the Anthropocene

In the evolutions induced by the Anthropocene, sea level rise is the most rapid driver of change for the protected areas located on low accumulation and dyked coastlines behind which brackish to freshwater marshes have been able to develop. The strategies that can be envisaged then have the choice between resisting, undergoing or adapting to these changes, as advocated in France in the national

6 For a critique of this patrimonial posture, see the analysis of "patrimoniomania" by Génot (2010).

strategy for coastline management, even if resisting seems a less and less conceivable horizon for technical and financial reasons. In fact, this situation forces us to question the baselines chosen until recently (especially contemporary ones) and requires us to imagine new baselines, depending on the possible effects of sea level rise (and the frequency of storms) and the heritage issues of biodiversity conservation. While these new baselines can be defined to adapt protected areas to sea level rise, they also increasingly integrate climate change mitigation services. The new European strategy for biodiversity by 2030 introduces the notion of carbon-rich ecosystems, which prioritizes not only biodiversity wealth, but also carbon sink services. The compositionalist approach to conservation is thus increasingly coupled with, if not replaced by, a functionalist approach. Second, the redefinition of baselines is now necessarily thought of not only on the scale of the perimeter of the protected area, but on the scale of the entire territory in which the latter is inserted, which obliges us to negotiate them with the territory's stakeholders: local communities, associations, economic stakeholders (especially farmers), inhabitants, tourism sector, etc. These new baselines are thus designed within the framework of a territorial recomposition beyond the protected areas or by redrawing their outlines. They are ultimately based on "win–win" strategies that intend to support and control the effects of sea level rise, both by limiting and controlling the risks of marine submersion, choosing the natural environments "to be lost" and those that can be "gained" in a biodiverse horizon, and by trying to improve carbon sink services. The construction of such scenarios can be accompanied by cost–benefit analyses based on the tools of environmental economics. These strategies can be called various things, but their objectives remain more or less the same, whether it be adaptive management, which re-evaluates the choices and objectives of ecosystem management in light of climate change, managed realignment strategy (Esteves 2014), which aims to allow for the retreat or translocation of the most valuable habitats and the development of new ones, or flexible shoreline management, which consists of using "nature-based solutions" to regulate the effects of sea level rise. In other words, these adaptive scenarios attempt to imagine fewer dykes and diked marshes and more "natural" mudflats and shores to mitigate the risks of marine submersion, and which are at the same time natural environments with high levels of biodiversity and that are rich in carbon. Figure 9.2 proposes scenarios for the evolution of protected areas located immediately behind the coastline. It does not aim to be exhaustive, but it includes some of the most important elements: new habitats and the variation of the protected perimeter, depending on the shoreline management chosen (hold the line, managed realignment to laissez-faire).

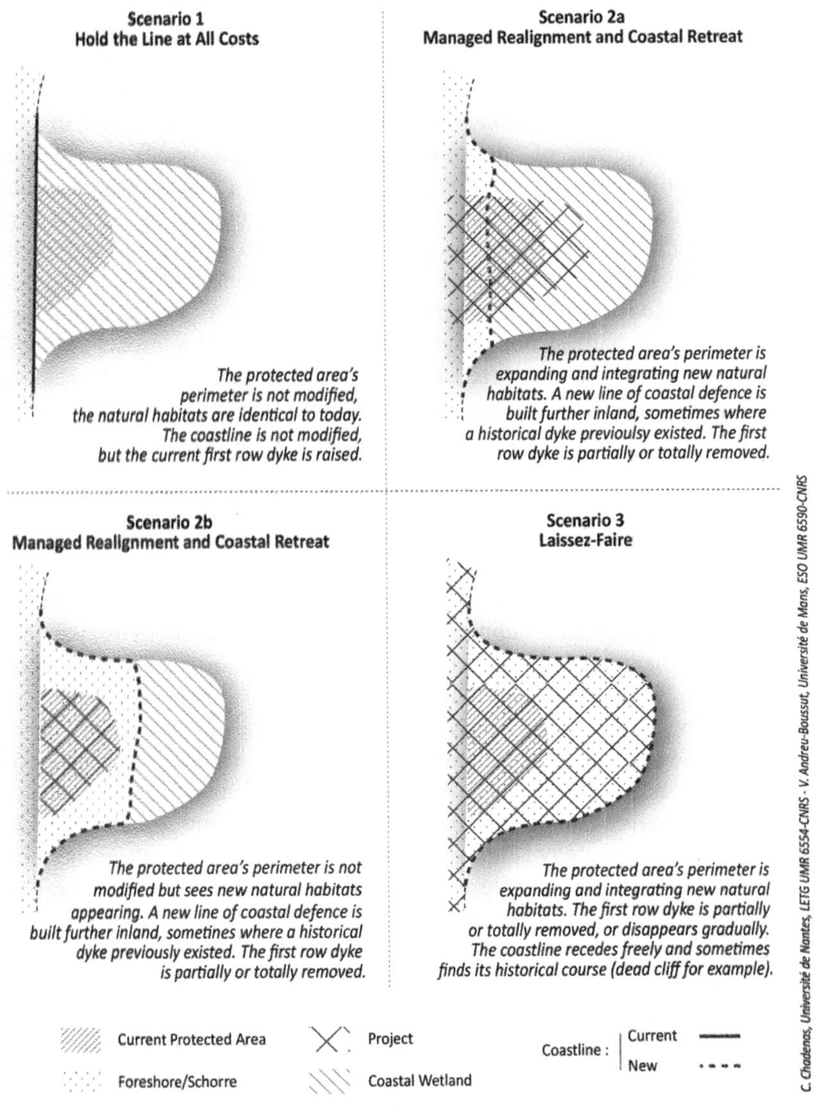

Figure 9.2. *Sea level rise adaptation scenarios for protected areas located on dyked coastlines. For a color version of this figure, see www.iste.co.uk/godet/baselines.zip*

De-polderizations have thus become frequent on the scale of the European coastline (Goeldner-Gianella 2009), even if Germany, the United Kingdom and the Netherlands have a head start. However, the latter can take different forms depending on the degree of sea backflow on the former polders, and it is thus possible to distinguish between controlled de-polderizations through tidal gates, partial de-polderizations through accidental or voluntary breaches, and complete de-polderizations through the removal of dikes (Goeldner-Gianella 2007). Formerly, dyked marshes thus give way to mudflats and salt marshes, which then emerge spontaneously and progressively by passive restoration or can be recreated by ecological engineering interventions. Although such operations are rarely the subject of open conflict, the weight of social representations favorable to polders and a strong attachment to their cultural landscape values constitutes a permanent brake (Goeldner-Gianella 2012), particularly in the Netherlands and Germany where the culture of the dike is well anchored in local societies. The extent and diversity of de-polderizations on the scale of the northwestern European coastline has been particularly documented by Lydie Goeldner-Gianella (see in particular Goeldner-Gianella (2007)), who proposes to distinguish between them according to their assumed objective: coastal defense, biodiversity, compensatory or nature-based tourism development. Nevertheless, it seems that in the most recent operations, these objectives are increasingly combined and conjugated, like the de-polderization carried out from 2014 onwards at Langwarder Groden in the Niedersachsen National Park (Germany). The restoration of 140 hectares of saltwater environments (mudflats, salt marshes) by creating a breach in the summer dike of a polder built in the 1930s was initially thought of as a compensatory measure for the development of the Wilhelmshaven container terminal. Nevertheless, this de-polderization is also a project of biodiversification of a former agricultural polder, and its long territorial negotiation also allows us to add a tourism purpose to it, which convinced inhabitants and local officials to see the project as an opportunity for redevelopment through nature-based tourism and an argument in favor of new tourism marketing for their territory.

These new baselines, adapted and adjusted to the Anthropocene context, are not guided specifically by a return to the past, but directed by a search for more naturalness in ecological processes and oriented by the socio-economic uses and political stakes of the territories. They constitute a "new nature" (Jansen *et al.* 2017) that can be guided and oriented by ecological engineering interventions (Jansen *et al.* 2016). On the European coast, this reorientation of baselines was initially carried out in the 1990s in the Netherlands through the doctrine of Nature Development (Barraud 2021). This doctrine served as a conceptual framework for the redefinition of nature conservation strategies on a national scale by proposing a new combination of protection and ecological restoration through the notion of natural infrastructure (Baerselman and Vera 1995). These baselines can also be considered as "novel ecosystems", in the sense that they may deviate from historically known patterns based on global change and local management decisions (Hobbs *et al.* 2013). They are neither "wild" nor purely artificial and lie on a continuum between these two extreme baselines (Hobbs *et al.* 2006). They are the best possible combination of maximizing beneficial changes and minimizing less beneficial transformations (Hobbs *et al.* 2006).

While this is indeed a new horizon for protected areas' management, it is also open to criticism insofar as it ultimately constitutes a technicist headlong rush, through geo-engineering or ecological engineering, and marks the pursuit of a more or less advanced control of the dynamic of ecosystem evolution (Schlosberg 2017), giving "birth to a new anthropocentrism that would value the technical power of humans over evolutionary ecological processes, a technical power that scientists and ecological engineers would hold" (Blandin 2009, p. 84). The example of the Wadden Sea, listed as a UNESCO World Heritage Site in 2009, is exemplary in this respect. Since the creation of the first reserves in the Netherlands and Denmark in the 1970s and of the German National Parks in the following decade, the management of this vast coastal area has remained rooted and faithful to the precept of "Let Nature Be Nature!" However, the mudflats and shores that make up the ecological value of the Wadden Sea are now threatened by sea level rise. This process of coastal squeezing is simple; it is now expected that the sea level can experience a faster rate than that of sedimentation of mudflats located between the islands and the rigid line that constitutes the largely dyked coastline, which does not offer any possibility of retreat to the mainland to all these intertidal environments. The German ecologist Karsten Reise, an emblematic figure of the Wadden Islands, proposes a management scenario with two complementary intervention methods. On the one hand, it is a matter of accentuating de-polderizations behind the dikes in order to recreate mudflats and salt marshes to compensate for the areas that would be lost between the islands and the shoreline. And, on the other hand, it is a matter of regularly and sufficiently supplying each sediment cell with sediment pumped from the sea, in order to accelerate the sedimentation rate of mudflats in the face of rising sea levels (Reise 2013). Thus, a sediment pump on an unprecedented scale compensates for the impossibility of this immense ecosystem to operate a strategic retreat towards the continent and behind the dikes. Such a scenario has already been proposed and implemented by the state of Schleswig-Holstein in its Wadden Sea Strategy 2100[7] since 2015.

9.6. The baselines of novel ecosystems in free evolution

In the last decade, a new conservation approach has attempted to promote the free evolution of ecosystems (see Chapter 6). This new posture proposes practices of non-intervention that articulate responses to two types of criticism regularly leveled

7 Ministerium für Energiewende, Landwirtschaft, Umwelt und ländliche Räume des Landes Schleswig-Holstein (2015). *Strategie für das Wattenmeer 2100*. Kiel, p. 85.

at "orthodox" nature conservation. First, it is a matter of turning our back on historicist or contemporary baselines that tend to lose their meaning and feasibility in the context of global changes induced by the Anthropocene. New baselines can now be produced, not only in terms of current or past biodiversity, but in terms of the current and future trajectories of ecosystems, in an evolutionary ethic (Blandin 2009). It is then a matter of discarding a fixist vision of the objectives assigned to ecosystem management to embrace an "open-ended" conservation (Lorimer 2015), *i.e.*, without a definitive horizon. It is then a matter of abandoning the steering of ecosystems by biodiversity and allowing a certain spontaneity of natural systems to express itself and allow for more unexpected events. Some authors therefore see the possibility of expressing greater naturalness in ecological processes (Schnitzler *et al.* 2008a, 2008b; Génot 2010). However, free evolution does not necessarily imply the abandonment of baselines in conservation projects. It is generally more about drawing a new desired state (Blandin 2009) and allowing the existence and evolution of "novel ecosystems" produced by the Anthropocene (Clement 2021). These can be interpreted as tools for thinking about dynamic and evolving ecological systems in a world marked by anthropogenic legacies and climate change (Marris *et al.* (2013), in Hobbs *et al.* (2013)). These "novel ecosystems" are all potential new ecological assemblages that erase the deep, artificial and inoperative dichotomy of thought between nature and culture (Monbiot 2014; Lorimer 2015) when the society–nature hybridity of our geographic spaces deserves central recognition (Whatmore 2002).

Until now, experimentation with free evolution has too often been a default choice rather than a real conservation strategy. In many cases, the free evolution of ecosystems arises by "accident" due to the lack of management documents in place or the lack of means to invest in the management of some protected areas. In the example of the Sainte-Lucie regional nature reserve, created in 2009, the former salt marshes saw spontaneous vegetation establish itself due to a lack of hydraulic maintenance for about 20 years before the managers had the opportunity to build their management plan from 2014 (Andreu-Boussut 2016; Pala *et al.* 2017). In this case, non-intervention allowed for the development of European-scale heritage habitats, a principle that guided the management plan in the choice not to re-water the northern part of the former salt flat. In contrast, in the early 2000s, on the island of Langeoog, the Niedersachsen National Park in Germany decided to remove the summer dyke built on the south side of the island, leaving the old polder to the tidal regime. The evolution of the habitats shows the natural restoration of a salt marsh, even if the plant successions seem slower than in other identical ecosystems, as they are still influenced by the former drainage system (Barkowski *et al.* 2009). Although managers have chosen to leave the ecosystem free to evolve, they are nevertheless attempting to accelerate the removal of small canals and ditches through ad hoc development to facilitate their sedimentation. Thus, allowing an ecosystem to express a new trajectory can be accompanied by initial interventions.

The choice of free evolution thus increasingly appears as a possible strategic horizon for some protected areas, when local contexts allow it. Since 2016, the Normandy Region has initiated a regional program of free-evolution spaces (*programme régional d'espaces en libre-évolution*, PRELE), whose network of protected areas is experimenting with non-intervention, with the dual aim of capitalizing on the knowledge acquired through the observation of ecological processes and producing a positive view of this new nature. European NGOs are also developing strategies favorable to free evolution, like the European Wilderness Society, whose Wild Coast label promotes protected areas with the "wildest" and least controlled ecological dynamics, such as the Jasmund National Park on the Baltic Sea in Germany or the Archipelago National Park in Finland. Nevertheless, such initiatives remain slowed down by the reluctance to control ecosystems less, which remains the technical and regulatory framework that protected areas are subject to. There is also a great risk of seeing, in non-intervention, a possibility of delegitimizing the work undertaken for decades by managers and the technical and financial means invested in a very "active" nature conservation.

9.7. Conclusion: the Anthropocene at the origin of new baselines for Littoral Protected Areas

While the Anthropocene is shaking up the ethical foundations of our relationship with nature insofar as its formulation acknowledges that there is no longer a nature untouched by human impact, this concept is also fueling the ongoing renewal of nature conservation practices and discourses on many coastlines. Until now, the creation of protected areas has met the objective of protecting and conserving a biodiversity threatened by anthropic pressures by maintaining environments in their current state, or by improving their ecological functioning, the difficulty, in many cases, being to know which state to refer to. The typology proposed here thus makes it possible to better understand the diversity of choices made, which are largely dependent on the interpretation and room for maneuver of managers. However, it is clear that sea level rise seems to be reshuffling the cards for nature conservation, especially for protected areas that are vulnerable to erosion and flooding and whose futures are at stake. Beyond the ecological issues, society will have to decide which kind of nature we want for the future in this unprecedented context, to borrow the title of Lévêque and Van Der Leeuw (2004). The Anthropocene now upsets the possibility of choosing baselines in protected areas that can no longer be considered as bubbles outside of time and global changes that affect both climate and biodiversity. It calls into question the possibilities of going back to highly contested ecological restorations (Alagona *et al.* 2012) and imposes the need to carefully and collectively decide from one site to another "the natures" that protected areas have the duty to conserve in the future.

In this new Anthropocene context, Wapner (2010) thus observes a growing divide among North American conservationists between a posture that assumes a "dream of domination" of nature, supported by improved knowledge and techniques and on ecosystem services that can be put on the financial market, and a posture that retreats to a "naturalist dream" and thus legitimizes renaturation projects based on choices to return natural spaces to previous premodern or prehistoric states. Distant as they may be, these two postures nonetheless bear a common double foundation in that they refuel and sustain a deep dichotomy between nature and culture (Lorimer 2015). But the Anthropocene also invites us to step outside this debate and brings up new registers of mobilizable baselines that do not erase the importance of knowledge of the historical trajectories of ecosystems; history thus remains a largely indispensable guide (Higgs *et al.* 2014). New approaches to nature conservation are developing in order to escape the "territorial trap" into which it has often fallen (Agnew in Lorimer (2015)), confined to fixed perimeters of protection that are rarely modified, and imprisoned in conservation management practices that have often forgotten the eminently dynamic character of ecosystems. Now, several questions are being asked: what will remain to be protected within the current perimeters of protected areas whose coastline position is bound to change rapidly? Should the protection perimeters be moved so that the natural habitats protected today will still be protected in the future? How much landscape and ecosystem spatial evolution are we willing to accept?

10

Baselines and French Forests

10.1. By way of introduction: "the legendary virgin forest of Doussard"

In the present state forest of Combe d'Ire (Savoie), located in the municipality of Doussard, France, there are two cantons of forest certified as virgin by the forest rangers of the 19th century.

> The woods are of fir, beech, elm and plane[1] species. The trees of these virgin forests are generally of a prodigious size: there are some that have seven to eight meters in circumference. In the thickness of the forests, the height of the trees is 45 meters and one commonly finds some with three meters, 50 centimeters circumference.

On July 3, 1856, four years before the attachment of the Duchy of Savoie to France, the Intendant of Albertville proceeded with the sale of more than 250 ha of these virgin forests in favor of "Dame Franceline Schmidt, widow of Mr. Jérôme Perravex d'Alex", forests that were exploited by clear cutting. A banquet was even given for the occasion. It was Ernest Guinier, director of the French National School of Water and Forests (*École nationale des Eaux et Forêts*), who related this event, in a conference given in Annecy on February 17, 1901 (Guinier 1905). Guinier added that "it is that indeed there are no more virgin forests in France. At least there are no more forest districts which, having been abandoned to the action of nature and not having been the object of any exploitation, are populated with trees of such strong dimensions".

Chapter written by Damien MARAGE.

1 Norway maple *Acer platanoides* L. 1753.

What seemed to strike the author and his contemporaries was the monumental size of these trees: diameters from 223 to 255 cm and an average of 111 cm for a height of 45 m; trees with an average unit volume of more than 100 m^3! Let us note in passing the indigence of the French lowland forests with regard to this "virgin" state: in 1856, the French forest surface was in the early stages of its great restoration (Mather *et al.* 1999). All the forests of the plains, coast and hills had been cut down, grazed and burdened with multiple rights of use (litter, dead wood, bark, etc.): they had indeed lost their virginity (Bépoix and Richard 2019). We would have to go much further east, to Poland, Hungary or Romania, to find forests with such large trees (Korpel 1995; Giurgiu *et al.* 2001; Bobiec *et al.* 2005).

But let us imagine for a moment that this sale had not taken place, that the "virgin" forest of Doussard had not been exploited. Would it be a baseline? But for whom? (i) the forest manager/owner, (ii) the scientist, (iii) the citizen who desires the "wild"? A baseline for what? (i) an unattainable absolute? (ii) an open-air museum? (ii) a target, a guide for the forest manager?

> [...] so humans who seek to discover the laws of nature desire places where they can study these laws. From this came the notion of reserves, national parks, and nature protection [...] but it is certain that if humans want to create a reasonable artificial dynamic, they must begin by knowing the natural dynamic.

Gaussen (1951) proposed this at the time when the commission for reserves in the state forest of Fontainebleau was being set up (Guinier 1950). This commission, from 1953 to 1973, recognized and integrated a little more than 600 ha as biological reserves in the successive forestry developments of Fontainebleau. Gradually and concomitantly, a network of about 120 forest biological reserves was established in French forests (Drapier 2005). But it was in 1998 that the instruction of the French National Forestry Office (*Office national des forêts* (ONF)) detailed the philosophy and conditions for the creation of integral biological reserves (ONF 1998). Two main arguments were put forward: (i) the consideration of biodiversity in forest management, justifying the creation of these reserves as **conservatoires of natural evolutionary processes** (Rameau 1995); (ii) the need to have **baselines** of sufficient size to "accumulate knowledge that can provide a solid foundation for certain silvicultural inflections required by changes in the ecological, economic, social and cultural context" (Barthod and Trouvilliez 2002), objectives in direct line with Gaussen's remarks. Forest managers lack reference values for the attributes of these unique forest areas. Such benchmarks could be derived from characteristics observed in unmanaged forests and/or reserves with comparable mesological conditions; for example, the number of large standing dead trees in a network of forest reserves could be used as a comparative element of forest management in similar stand types. In addition to studies conducted for decades in reserves in Fontainebleau and La Massane forests (Pontailler *et al.* 1997; Mountford 2002; Christensen *et al.* 2005; Burrascano *et al.* 2013), from the 1990s onwards, numerous works, first of inventories and then of dendrometric, dendrological and functional characterizations of forests with natural characteristics, followed one another (Mortier 1990; Greslier 1994; Closset-Koop 2004; Gilg 2005; Marage 2005, 2006), until the recent implementation of systematic data collection devices (Paillet *et al.* 2015).

Whatever their destination or use, these parcels, these cantons, these "virgin" massifs, arouse curiosity. Let us stop, first of all, on the structure, composition and functioning which characterize these so singular forest stands, commonly considered as baselines for the forest environments.

Some forest definitions (Da Lage and Métaillé 2000; Bastien and Gauberville 2011; FAO 2020)

Virgin forest: forest formed exclusively by natural factors and in which biological processes occur without any direct or indirect human influence.

Natural forest – two views:

1) Forests whose stands are made up of species indigenous to the area under consideration, established by natural regeneration.

2) Forests whose structure shows direct or indirect anthropic influences, but which retains, to a large extent, their natural structure. They are made up of tree species corresponding to the (forest) site.

Subnatural forest: a forest that has been managed by humans, but has retained or regained (after management has ceased) a composition and structure close to that of a natural forest (according to definition two above). The synonym of "subnatural" can be "natural".

Primary forest – two views:

1) Naturally regenerated forest of native species where no evidence of human activity is clearly visible and ecological processes are not significantly disturbed.

2) Generally synonymous with virgin forest. Forest that is or has returned to the mature, climax phase. Forest resulting from a primary plant succession.

Box 10.1. *Some forest definitions*

10.2. Forestry "cardiology"; forestry "systoles" and "diastoles"

Like cardiology, natural forest dynamics, called sylvigenesis (Oldeman 1990), has two main phases. The first is a phase of contraction of creative energies. It is an anabolic phase (production of living matter) under the dependence of autotrophic plant populations (capable of generating their own organic matter), which organize and accumulate biomass. This phase can be described as "systole", during which the forest grows. The second phase is that of an expansion, a catabolic phase (the tissues are transformed into energy) that puts heterotrophic populations into action (which cannot synthesize the elements constituting the organic substances by themselves), which disorganize and mineralize the organic matter. The forest can then be seen as in the "diastole" phase: it ages and then dies. This temporal dynamic takes place at three interlocking spatial scales:

(i) that of the tree: this is the space occupied by a tree during its lifetime. This "local" biotope is called the **ecotope** (Oldeman 1974, 1990). The processes at this scale are directly related to the architectural model specific to each tree. Among the 24 models that exist worldwide, let us quote Massart for the Fir, Rauh for Pines, Troll for Beech, for the trees of our world's temperate regions (Oldeman 1990). Throughout its growth, the spatial relationships of the tree with its closest neighbors change, firstly during juvenile growth and then as the crowns compete. The development of the twig, self-shading and various reiterations during its lifetime modify microclimatic conditions and, in doing so, the herbaceous vegetation. It is on this scale that the "out of the ordinary", "never seen before" dimensions of monumental trees are sometimes qualified, expressing their biological maximum, including for pioneer species (*Salix, Alnus, Betula, Populus, etc.*). The interweaving of ecotopes creates the structure of the forest stand;

(ii) that of inter and/or intraspecific tree collectives. Each surface on which at a given moment the regeneration of the vegetation is observed and on which the structure and ecophysiological processes are ordered by a group of trees. This constitutes an **eco-unit,** a regeneration unit, until the next disturbance. The death of the stems, breaking the forest cover, initiates the establishment of a new eco-unit. The authors distinguish eco-units according to their morphology (Oldeman 1974, 1990), the dominance of functional groups of trees (pioneer, post-pioneer, dryad; Rameau 1987) and the evolution of biomass and necromass[2] (innovation, aggradation, biostatic and senescence; Koop and Hilgen 1987; Oldeman 1990);

2 Amount of standing dead trees and their debris on the ground.

(iii) that of a massif, of an entire forest. The whole of the eco-units constitutes the **sylvatic mosaic**. The surface must at least be constituted of the various eco-units since the phase of innovation until the phase of senescence. This minimum surface varies according to the types of forest ecosystems, from 30 to 50 ha at the collinean level to several hundreds of ha for the subalpine sandstones (Korpel 1995).

It is all of these spatio-temporal processes which form the sylvigenesis, dictated by the regime of local disturbances (fires, storms, frosts, insect pests, etc.). The characterization of this sylvatic mosaic allows the **sylvigenetic maturity** of the forest to be determined. In Europe, beech–fir forests with spruce in the alpine zone show the highest sylvigenetic maturity (Rameau 1987; Korpel 1995; Govedar *et al.* 2019).

10.2.1. *Forest systoles*

The forest is the center of a fundamental process: regeneration. This rejuvenation phase is caused by the more or less sudden and extensive separation of the canopy during a windfall (exogenous disturbance), or simply the death of a tree (endogenous disturbance). Watt (1947) made the first description for the forests of temperate Europe:

> To the three phases recognized – Bare, Oxalis, Rubus – a fourth should be added, the gap phase, to which regeneration is confined because it is excluded from other phases. At any given place there is a cycle of change consisting of an upgraded series of phases in which there is a continual change in ecological structure, associated with changing age, rate of growth and density of the dominant trees and correlated with changes in the field layer, and a downgrade of the dying, dead and rotting stems and the vegetation of the gap.

Jones (1945) noted that regular stand structures are not uncommon, with fire or storms being the primary agents of complete regeneration over large areas. The rate of canopy closure depends on the size of the gap, the pre-existing condition of the understory and the soil seed bank. During this reorganization process, the species that move in and colonize this opening guide the dynamics. Certain taxonomic groups are dependent on this phase, such as certain ground-nesting birds (Woodcock or European Nightjar when the gap is large), numerous lepidopterans taking advantage of a flowering herbaceous carpet, small mammals such as voles and shrews and, of course, large herbivorous mammals such as deer. In beech–fir–pasture forests in the Vosges, Bernard (2018) highlights a quasi-inversion of species between fir and spruce in the regeneration stratum, benefiting the spruce, when they are subjected to strong browsing pressure, underlining the influence of large ungulates in interaction with climatic conditions on the dynamics of forest regeneration.

10.2.2. *Forest diastoles*

Two types of material feed the forest diastole (aging): (i) litter, which has an annual dynamic in our latitudes and (ii) necromass, which has an irregular dynamic due to long-term processes (Pontailler *et al.* 1997; André 1997). A complex food web develops, mainly composed of insects, arachnids, crustaceans, mollusks, fungi (see Figure 10.1(c)), micro-organisms, certain guilds of birds (*e.g.*, woodpeckers; see Figure 10.1(c)) or mammals such as bats (such as the Common Noctule, Barbastelle or Bechstein's bat). All of these organisms, which depend on dead or dying wood for part of their life history, form the **saproxylic complex** (Harmon *et al.* 1986). These organisms operate in three successive stages: (i) colonization (*e.g.*, mite penetration of a trunk); (ii) decomposition; and (iii) humification. When only lignin and hemicellulose are left as an energetic substrate, the length of the food web inevitably increases. Thus, a larva of Lucanus kite takes three to five years to complete its metamorphosis. Bobiec *et al.* (2005) estimate the decomposition process of dead wood to be between five and 100 years, depending on the species, the size of the wood and its contact with the ground.

The importance of the saproxylic complex goes beyond the presence and abundance of certain functional groups, but leads to certain ecological processes occurring. For example, the decomposition of very large diameter trunks (> 150 cm) on the ground promotes the natural regeneration of trees, either by limiting browsing or because of the greater organo-mineral richness of the contiguous soil. This is the nurse effect (see Figure 10.1(b); Bobiec *et al.* (2005)). When decaying trunks are still standing, forms of epiphytism occur for the same reasons and contribute to the healing of the gap through seed rain (see Figure 10.1(a); Marage *et al.* (2006)).

Figure 10.1. *Examples of epiphytism with Rubus ideaus and Epilobium angustifolium on decomposing beech (a); "nurse" effect on the decomposed oak trunk (b); and tinder carpophores and woodpecker borings on Ash (c) (source: author's own photos). For a color version of this figure, see www.iste.co.uk/godet/baselines.zip*

10.2.3. *Phases of silvigenesis: when only certain attributes are retained*

According to the evolution of biomass and necromass, the rhythm of sylvigenesis is divided into four phases: innovation; aggradation; biostatic and senescence (see Table 10.1). The innovation phase allows the soil seed bank to express itself, herbaceous vegetation dominates, as do seedlings and some trees of the present. During the aggradation phase, the canopy closes, inter- and intraspecific competition is exacerbated, the herbaceous carpet tends to disappear and the vertical structure is unistratified. Then, the trees of the present dominate, expressing their maximum biological potential; the biomass values are then maximal. For the lowland beech forest, the volumes can reach 500–600 m/ha^3; in the mountains, 900–1,000 m/ha^3 in the beech–fir forests. For internal (maximum life span reached) or external reasons (fire, storms, floods, etc.), the biostatic state is modified, and the upper stratum opens up. The volume and density of standing dead trees increase, followed by that of dead wood on the ground, with volumes often exceeding 100 m/ha^3. This is the senescence phase. In Fontainebleau, in the Tillaie reserve, the volume of dead beech on the ground was between 107 and 256 m/ha^3, just after the

1999 storm (Mountford 2002). In the Massane beech–hornbeam forest (Pyrénées Orientales), the volume of dead wood was about 35 m/ha^3 (Christensen *et al.* 2005). In the beech–fir forests of Bois du Chapitre (Hautes-Alpes), the average volume of standing dead wood was estimated at 202 m/ha^3 (Marage and Lemperiere 2005). On a national scale, Paillet *et al.* (2015) estimated an average total volume of 41.7 m/ha^3 for lowland beech–oak forests (*e.g.*, Fontainebleau, Chizé, Citeaux), compared to 60.2 m/ha^3 for mountain forests (Vosges, Jura, Alpes).

The biostatic and senescence phases have all the attributes of old-growth stands (Harmon *et al.* 1986; Gilg 2005). Because of the sometimes magisterial size of some trees, the considerable volume of the stands, the importance of the saproxylic complex and the perception of a certain homeostasis (Oldeman 1990; Haguenauer 1991), these are two emblematic phases "appreciated" by scientists, forest rangers, naturalists and even citizens (Brahic *et al.* 2016).

Eco-unit	Duration (years)	Surface (m^2)	Saproxylic complex	Biological and ecological processes
Innovation	2–40	10–10^{10}	+	Reiteration Mycorrhization Herbivory
Aggradation	40–100	f(innovation)	-	-
Biostatic	>100	>500	++	Herbivory
Senescence	5–100	f(biostatic)	+++	Nurse-effect

Table 10.1. *Main characteristics of the phases of the silvigenetic cycle in France (from Lemée 1989; Mortier 1990; André 1998; Schnitzler-Lenoble 2002; Marage 2005, 2006)*

All of the French studies reveal a great variety of sylvigenesis, among which a state of equilibrium seems to be a special case. For Cordonnier (2004), this does not mean that dynamic equilibria do not exist, but only that they should not constitute a privileged reading window for defining a baseline. Equilibrium regimes can be transient and very different from the dynamic regime. The equilibrium state is convenient because it allows a baseline to be defined without ambiguity. The baseline does not necessarily correspond to an equilibrium state, but rather to a state (or a dynamic) considered as "normal" or desirable: it is the target to be reached.

The authors therefore draw on forests described as "subnatural" as baselines, forests where management has been abandoned for more or less a long time. If for species of the saproxylic complex the subnatural forest can be a baseline, what about other species or functional groups of the systolic phase? Is their natural state so natural that traces of past silvicultural orientations can be found?

Timbal and Maizeret (1998) ask, with regard to the Landes forest, whether the present forest, which was "created" from scratch in the 18th and 19th centuries, should be considered as the baseline, or whether it should be the more or less humid moorland ecocomplex that dominated until the 19th century, or the forest that was present at the time of the post-glacial climatic optimum? Becker (1972) has shown that in the Charmes forest (Vosges), the hydromorphy of the soils was not the cause of the degradation of the stands, but rather the consequence of past abusive logging, as has Georges Plaisance in the Chaux forest (Jura). Houzard (1986) has clearly shown the influence of past silvicultural treatments, particularly short rotation coppicing, on the current state of the soils in the forests of Lower Normandy. The development of surface podzolization phenomena is the result of soil impoverishment following the overexploitation of the forest environment. Thus, interpreting the logic of forest dynamics in light of historical events has become crucial (Rameau 1987; Simon 1999).

10.3. The baseline of French forests examined through the lens of historical ecology

10.3.1. *The vicissitudes of forest cover since the Holocene*

Using a range of paleo-environmental tools, archaeology and ancient textual sources, we can "go back in time". For example, based on the work of Mather *et al.* (1999), Kaplan *et al.* (2009) estimated forest cover rates in Europe using retrospective modeling of demography, climatic conditions and land use. Around 3,000 years BP, most of the European continent still had a forest cover of around 80% in areas that were still mainly occupied by non-sedentary prehistoric societies. For France, the simulation gives 79% cover at 3,000 years BP. During the Early Bronze Age, the use of alder, ash and hazel coppice is attested for fishing in the Bay of Mont Saint-Michel (Bernard *et al.* 2012). In the Moselle forests, Etienne *et al.* (2013) show that the pollen rain, from the Iron Age to the Gallo-Roman period, testifies to open landscapes, surely for agro-pastoral use, in a forest matrix that tended to decrease. This is confirmed by Kaplan *et al.* (2009), with a forest cover of about 46% during this same period.

Using the REVEALS model, Robert *et al.* (2018) were able to show that the French Atlantic coast has always been much less densely forested than central Europe. Forest extent began to decline shortly after ~6,000 BP, when Neolithic agriculture became established throughout Europe at these mid-latitudes. Pollen-based reconstructions indicate that the conversion of forests to agricultural land became visible on a subcontinental scale during the late Neolithic and Bronze Ages, in the late 5th and 4th millennia BP (Jouffroy-Bapicot *et al.* 2013). Forest loss was already well advanced by 3,000 years BP, with REVEALS estimates for Europe indicating a drop of about 63% from the mid-Holocene forest maximum (Roberts *et al.* 2018). All of these authors confirm that the anthropogenic impact on Europe's temperate hardwood forests was already significant before the Iron Age.

Moreover, Alpine forests were not spared: Conedera *et al.* (2017) and Dietre *et al.* (2020), combining palynology, sedimentology, anthracology and archaeological excavations, have shown that even very remote forest areas (Italian–Swiss border in the Silvreta massif) were exploited from 4,300 years BP onwards. This supports the archaeological interpretation of Neolithic sites, where all archeological, archeozoological and archeobotanical information pleads in favor of the presence of seasonal hunter–gatherer camps using wild animal and plant resources until 5,000 years BP (Dietre *et al.* 2020).

Everywhere, the millennial imprint of humans and their activities (grazing, firewood burning, etc.) has left its mark on the forest cover and more or less profoundly modified its composition (Bépoix and Richard 2019). Some species have benefited from forest exploitation or the abandonment of previously cleared areas: the progression of beech, treated as coppice, at the expense of fir to produce firewood in the Massif-Central, the Alps and the Pyrenees; the massive recolonization by spruce in the Alps and the Jura, by Scots pine in the Southern Alps, the Massif Central, the Pyrenees with the abandonment; on the hillside, hornbeam and more heliophilous oaks (*Q. robur, humilis, pyrenaica*) at the expense of beech and sessile oak by maintaining the coppice-subwood treatment; and finally, the extension of cork oak and especially holm oak at the expense of downy oak in the Mediterranean region. Placing an absolute baseline and sticking to it is therefore historical nonsense: under the paradigm of the balance of nature, it is a well-known and denounced fixist vision (Rameau 1987; Marage 2018). But why did some authors still want to present a map of Europe's primary forests in 2018 (Sabatini *et al.* 2018)?

10.3.2. *Baselines: compasses or daymarks?*

Two trends seem to be emerging: the supporters of the "compass" baseline to guide us towards a target to be reached and the lauders of the "daymarks" baseline to bring us back to the "harbor" of one of the past baselines.

For proponents of the "compass" approach, the baseline is a "desired optimal state", a state towards which we wish to strive (Carnino and Touroult 2010). Rather than favoring the collective expertise of "sources" to define this optimal desired state, as advocated by Maciejwski *et al.* (2016), a socio-ecological casuistry could be deployed here (Marage 2018; Marage *et al.* in press). In this framework, the conservation objective implies a form of collective deliberation on the target to be reached and recognizes that it is a societal choice and not a biological value that is imposed. Forest management then consists of steering, on the basis of scientific knowledge, and using techniques designed through that knowledge, the trajectories of forest dynamics so that they retain, recover or acquire socially chosen structural and functional characteristics and a degree of adaptability (Blandin 2014; Marage 2018).

For the "daymark" lauders (Machado *et al.* 2004; Scholes and Biggs 2005; Rodriguez *et al.* 2011; Winter 2012), the baseline is the pre-industrial one. Research on the naturalness of forests was initiated in the German-speaking countries of Central Europe using two German words: "*Naturnähe*", which evokes a comparison between the current state and a natural baseline, and "*Natürlichkeit*", which evokes the current state of the forest (Winter 2012). The baseline should represent the most natural forest with respect to growing conditions (environmental factors), such as climate, soil, exposure and elevation (Winter *et al.* 2010). Thus, in the context of implementing the Habitats-Wildlife Directive, Evans and Arvela (2011) state that favorable baseline values are key concepts in assessing conservation status. The reporting requires member states to identify **threshold values** for range and areas for habitat types, in order to assess whether the range and areas are large enough to conclude that the parameter is "favorable" or "unfavorable" and, if "unfavorable", whether the status is "inadequate" or "poor". The favorable baseline values must be based on **purely scientific grounds**: but what science are we talking about? As Chevassus-au-Louis *et al.* (2008) point out, on the empirical level, a close observation of ecosystems and their history has in fact led to highlighting the often contingent nature of their state at a given moment, resulting from historical phenomena (founding effects in biocenoses, more or less periodic "catastrophes", lasting impacts of past environments, etc.). As a result, an observed state of an ecosystem, even one considered "natural", could not be considered without caution as a "baseline" that should be preserved or, in case of human or natural disturbances, "restored".

In the management of protected areas, particularly forest-dominated nature reserves, the search for criteria and indicators from baseline forests allows for the definition of a theoretical state for each indicator (Gilg 2005; Winter 2010, Paillet *et al.* 2015). Analytical descriptions and the use of modeling allow us to propose scenarios to define baselines for forest habitats subject to management (Winter 2012) where baseline models are sought in "natural" forests, *i.e.*, untouched by anthropogenic action. According to Closset-Kopp (2004), “the aim is to find baseline sites that are as little influenced by human activities as possible and as similar as possible from a functional point of view”. These sites can be numerous, in order to gather all the attributes for a single measure of biological integrity, but must remain in close geographical contexts. In the Northern Vosges, the baseline values chosen were inspired by the Tillaie and Gros Fouteau reserves (Leroy 2006), following the guide for drafting management plans for nature reserves (Chiffaut 2006). This guide defines a theoretical baseline of a natural habitat as the state in which all the needs of the ecosystem are met, such as its ecological conditions, functioning and human activities, if any. Sketching the baseline of each habitat means defining an optimal or historical surface, a floristic composition, a balanced structure of the vegetation, the presence of natural regeneration, etc. These criteria and indicators are extracted from: (i) the state of the initial situation if it was considered satisfactory; (ii) the state observed in a preserved sector of the reserve (or outside), under equal ecological conditions, (iii) expert opinion; and (iv) the state described in the literature, for example, in the Natura 2000 habitat notebooks. Sometimes, the baseline is quasi-contemporary, as, for example, the alluvial forests of the Rhine where, in the framework of the restoration of their functionality, the state of the river before the great rectification works of the 19th century were chosen.

Is the use of a pre-industrial baseline, implicitly considered as “desirable”, socially acceptable? Its proponents defend this approach by specifying that it is not a desirable state, but a baseline that makes it possible to evaluate the current state by comparison and set political objectives in relation to past states, states taken arbitrarily. This is the case, outside of France, in Quebec, where baseline portraits are established as a basis for ecosystem-based forest management. The Sustainable Forest Development Act defines ecosystem-based management as “management that consists of ensuring the maintenance of biodiversity and the viability of ecosystems by reducing the differences between the managed forest and the natural forest”. In the reference manual for ecosystem-based forest management (Guay *et al.* 2013), the implementation of ecosystem-based management is used to reduce the gap between natural and managed landscapes. The pre-industrial forest is used as the natural forest baseline (Guay *et al.* 2013).

However, all the trajectories of forest ecosystems are inscribed in time. We cannot, like the meridians of Greenwich, Cordoba, Naples, Krakow or Paris, move our clocks forward or backward: the arrow of time runs uninterrupted through the course of any ecological system. Even if they are chosen arbitrarily, these different meridian systems are all correct in the absolute: you are found because you are able to give your position; it is only a question of reference to a time 0. Choosing your baseline meridian is arbitrary, but it is possible to convert this time difference between the original meridians. For the trajectories of forest ecosystems, some arbitrarily choose a baseline meridian, but do not have the "clocks" to set themselves and go back in time, and even less in the Anthropocene era.

10.4. French forests in the Anthropocene era: chosen or endured states?

10.4.1. *A state chosen according to a geography of conservation*

10.4.1.1. *The return of the wilderness because we choose it: wildlife reserves and free evolution*

> To make naturalness an objective, should all intervention be excluded? No more management of nature, therefore, but a laissez-faire attitude, letting it evolve. Are we sure, in doing so, that we are going in the right direction?

These were the questions asked by Barbault in 2008 during the *Réserves naturelles de France* conference. The fact is that free-evolution forestry projects are *flourishing,* such as the "*Forêts sauvages*" association, the *association pour la protection des animaux sauvages* (ASPAS), the *Groupement Forestier pour la sauvegarde des feuillus du Morvan*, numerous forestry plots of the *Conservatoires d'espaces naturels,* etc. The purpose of the non-intervention of forest plots is to have freely evolving reference points that lead to the evaluation of the effectiveness of more interventionist strategies. In the sense of Hobbs *et al.* (2006), these are "novel ecosystems", *i.e.*, ecosystems profoundly modified by human activities that are no longer managed and thus left to their spontaneous dynamics. These new ecosystems are the result of an often original combination of species. In addition to forest stands resulting from old silvicultural itineraries, RTM reforestation[3] abandoned several decades ago, recent forests resulting from agricultural abandonment fall within this field. Conservationists want to save, fight against, conserve certain habitats or species, etc. We rarely hear them say what these ecosystems should look like in 50 or 100 years. These stands left to evolve freely are part of a forest and agropastoral landscape mosaic. Free evolution seems, in the eyes of their defenders, to only have advantages: resilience, a pool of biodiversity, etc., which should not prevent us from analyzing the risks that these new ecosystems may pose to adjacent ecosystems, for example, with respect to invasive species (*e.g.*, *Rhododendron ponticum* in Killarney National Park in Ireland; Cross (1981); Gioria *et al.* (2018)) or increased probability of fire (Fernandez-Manjarres *et al.* 2018).

3 *Restauration des Terrains en Montagne* (mountain terrain restoration): the French Forest Service, which has attempted reforestation of badlands for erosion control since 1860.

10.4.1.2. *Adaptive management and ecological restoration of forests*

Adaptive management implies the systematic use of modeling, experimentation and monitoring to compare the effects of several management strategies: it is easy to understand that the length of forest cycles makes its implementation tricky. Nevertheless, designing management measures, as well as experiments that encourage learning and innovation have been practiced for decades by forest managers, such as plantation trials, seed orchards and the selection method in planted forests. The management of continuous cover forests is part of this perspective when it seeks to evaluate the ecological and socio-economic responses to management measures and involve social actors in the understanding and management of a multifunctional system (Bruciamacchie 2009). To do this, it uses Holling's conceptual apparatus based on an in-depth knowledge of hierarchical organization and adaptive cycles (Holling 2001; Folke *et al.* 2004). This management should lead to an increase in the adaptive capacity of forest socio-ecological systems.

Some habitats, degraded or simply fragile, require conservation or restoration actions. Ecological restoration implements intentional actions that initiate, support and promote the recovery of an ecosystem that has been degraded, damaged or destroyed, while respecting its health, integrity and sustainable management (Aronson *et al.* 2007). Ecological restoration aims for a baseline that is ecologically possible and achievable through human intervention. From the point of view of social acceptability, with climate change, a need for adaptation of native species is expected: will they adapt? Should they be helped to do so? The question is not only a scientific one, because the issues at stake today are those of territorial management, with the desire to be active in the future adaptation process and decide collectively. The reflection must be extended to social, economic and political issues. Indeed, proposals for assisted migration are multiplying (*e.g.*, the GIONO project in the Verdun forest, numerous trials by the *Société forestière de la Caisse des Dépôts et Consignations,* by independent forestry experts, etc.). For a long time, forest managers have been saying that they are adapting their forest stands to climate change. But what about the evolutionary processes: should they be forced and if so, how? Bringing trees from elsewhere to genetically mix and reinforce, or not, the potential of adaptation is a very concrete way to respond. However, current projects take risks, to the point of losing all "reference". It is necessary to accompany this approach with a broad local consultation and discussion with all stakeholders. But will it be possible to keep all of these processes "under control"?

10.4.2. *A state undergoing global changes*

In the state forest of Compiègne, the late cherry tree (*Prunus serotina*) did not ask the forest rangers for their opinion. Originally from North America, it was introduced in France around 1623–1629, and then planted in European forests for various purposes. Since the early 1960s, it has spread widely in the temperate forests of Western and Central Europe, particularly on well-drained, nutrient-poor soils, and has been recognized as one of the major forest invaders. It is able to out-compete native species in openings to form a low, closed carpet of small trees that impedes natural regeneration and impacts the functional diversity of understory plant communities (Chabrerie *et al.* 2010).

Atmospheric nitrogen deposition also changes the composition of the forest floor throughout the country, and this is not something that can be controlled. In Switzerland, in beech–fir forests in Aeschau and spruce forests in Bachtel, N deposition >25 kg/ha/yr exponentially increases the cover of bramble (*Rubus fruticosus*) (Belyazid *et al.* 2011). Recent increases in nitrophilic forest floor species due to increased rates of N deposition have been recorded across Europe. These include studies showing an increase in nitrophilous species in Dutch forests with deposition > 40 kg/ha/yr, an increase in nitrophilic species in fir and Scots pine forests in Germany with a deposition of 15–30 kg/ha/yr, a decrease in the frequency of many species and an increase in the frequency of nitrophilic species in the central plateau of Switzerland with a deposition of 30–40 kg/ha/yr and an increase in nitrophilic species in deciduous forests in eastern France with a deposition of 20–30 kg/ha/yr (Bobbink *et al.* 2010). Although other factors (*e.g.*, forest management) may alter species composition, together, these studies provide strong and consistent evidence that N deposition has a significant impact on European temperate forests.

Tree growth is also modified. Bontemps *et al.* (2012) concluded that there has been a recent divergence, with a decline in the growth of beech versus a dramatic increase in the growth of sessile oak, especially in the Grand-Est region. Moreover, the analysis of height trajectories shows a recent reduction in the competitiveness of common beech compared to that of sessile oak. Last but not least, changes in the climatic regime are leading to increased mortality in a large majority of trees (Taccoen *et al.* 2019). We are a generation of forest rangers, researchers and citizens who will observe, measure and feel these changes. We must prepare our minds for these changes. Trees are dying, entire massifs will surely disappear. We must prepare the local residents, the inhabitants of these territories and our fellow citizens for these changes. This will disrupt the landscape. Forest rangers are already suffering psychologically. A forest ranger, therefore an informed professional, told me of this suffering in 2019:

> I became a forest ranger to maintain the health of the forests, to see to the growth, the regeneration, the growth of the trees; I did not become a forest ranger to harvest dead trees! (when talking about common spruce trees).

We can say that this forest ranger is indeed a victim of solastalgia[4], in the words of Glenn Albrecht. Let us hope that we are not also victims of a "baseline shift", a term for changes in state that escape collective memory (see Chapter 4). This form of intergenerational amnesia, where the image we had of the state of an ecological system is the one we had at the beginning of our lives or careers, is the one that refers to our own experience (Blondel 2020). Are the managers of the present Combe d'Ire Biological Reserve aware that just over 100 years ago the brown bear was clawing at the beech trees? How did this great predator regulate the chamois populations in the Bauges? How did he modify the dynamics of regeneration? We will never know. In a few decades, will the hiker in the Vosges know that the capercaillie was parading near the great trail (long distance hike) in the Grand Ventron massif?

10.5. Baselines and French forests: illustrating our forests together

Let us conclude on the tutelary figure of Bachelard (1957). Let us listen to him telling us about his forest, and the poetics of the space:

> In the fields of poetic phenomenology that we study, there is an adjective of which the metaphysician of the imaginary must be wary: it is the **ancestral** adjective. To this adjective, in fact, corresponds a too fast valorization, often all verbal, never well watched, which makes us miss the direct character of the imaginary of the depths, even, in general, the psychology of the depths. The ancestral forest is then a cheap psychological transcendent. The ancestral forest is an image for children's books. If there is, with regard to this image, a phenomenological problem to be posed, it is to know for what actual reason, in virtue of what value of imagination in act, such an image seduces, speaks to us. A distant impregnation coming from the infinity of the ages is a free psychological hypothesis. Such a hypothesis would be an invitation to laziness if it were retained by a phenomenologist. As far as we are concerned, we believe ourselves obliged (sic) to establish the actuality of the archetypes. In any case,

4 A form of physical or existential distress caused by environmental change. The term captures the anxiety and grief of residents deprived of the comfort of home; to also capture their sense of helplessness in the face of environmental degradation (Albrecht 2006).

> the word ancestral, in the reign of the values of imagination, is a word to be explained; it is not an explanatory word.
>
> But who will tell us about the temporal dimension of the Forest? History is not enough. It is necessary to know how the Forest lives to its great age, why there are no young forests in the reign of the imaginary. For me, I only know how to contemplate the things in my country. I know how to live, Gaston Roupnel, the unforgettable friend, taught me, the dialectic of the country and wooded areas. In the vast world of the not-me, the not-me of the fields is not the same as the not-me of the forests. The forest is a before-me, a before-us. For the fields and meadows, my dreams and memories accompany them in all the times of ploughing and harvesting. When the dialectic of the me and the not-me is softened, I feel the meadows and fields with me, within-me, within-us. **But the forest reigns in the antecedent**. In such a wood that I know, my grandfather got lost. I was told about it, I have not forgotten it. It was a time when I did not live. My oldest memories are a hundred years old or a little older. This is my ancestral forest. And all the rest is literature.

There would be no salvation without the antecedent, without summoning the ancestral. But then how are the forests of the future composed? With what compass? First of all, forest ecosystems allow us to collectively reappropriate the times of reality without sinking into the descriptive or prescriptive, or getting lost in the search for a historical baseline. Secondly, they show that it is possible to propose a shared territorial and landscape governance made of a commonly desired future, which gives a concrete grip on the forests within these different temporalities back to the collectives (forest rangers, communities, citizens, etc.). In other words, to make our imaginary baselines a forestry policy in the territories because the creative power of our forestry imaginary is everyone's business.

11

How Can We Maintain Traditional Agro-Pastoral Landscapes?

11.1. Introduction

"In the theater of the fields[1], it was better before"! How beautiful they seemed indeed and so much more natural the gardened countryside of the first half of the 19th century, sublimated by the impressionists! How rich in fauna and flora must the agrarian landscapes have been before the rearrangements, the implementation of the PAC[2] and the industrialization of production systems! The facts seem indisputable. For more than 30 years, there have been countless studies demonstrating how the reduction of grasslands, the uprooting of hedgerows and the draining of marshes have contributed to the decline in biodiversity.

The transformation of landscape structures is thus the designated culprit, and it would then suffice to restore, replant, restore or renovate certain "traditional" agro-pastoral landscapes in order to renaturalize the French and European countryside. The "re" strategy is thus found at the heart of contemporary management logics (Gramond 2014), indicating the inevitable recourse to a past baseline to improve the present. The Society for Ecological Restoration defines ecological restoration as "the process of assisting in the recovery of an ecosystem that has been degraded, damaged, or destroyed". The state to be restored would then be the pre-degradation state, which is either a past configuration or a dynamic: the historical trajectory of an ecosystem before it was deviated by human disturbance (Balaguer *et al.* 2014).

Chapter written by David MONTEMBAULT.

1 In reference to the "*théâtre d'agriculture et mesnage des champs*" by Olivier de Serres (1600).

2 *Politique Agricole Commune*, French Common Agricultural Policy.

In all cases, the question of the baseline arises (Aronson *et al.* 1995). What are the traditional agrarian forms that deserve to be restored? At what point can human action be judged as degrading or deviant? To what period should we return to? An initial overview suggests that attention is primarily focused on sylvo-pastoral forms, which are less denatured than cultivated[3] areas. Since 2015, the PAC has encouraged, for example, the maintenance of permanent grasslands, moors and other rangelands or summer pastures, as well as ponds, hedges or isolated trees, as "Areas of Ecological Interest" justifying the "green payment". However, renaturation operations can also concern other spaces, with very variable dimensions. A quick assessment of the Natura 2000 network in metropolitan France, for example, shows that 12.9% of land areas are concerned, of which 15% are classified as "agricultural areas", 22% as "wetlands" and 26% as "moors and open environments"[4]. Knowing that agrarian systems include moors and wetlands, the identification of legacies appears complicated from the outset.

In order to try to see this more clearly, we will begin by redefining what an agrarian landscape is and how it evolves, in the systemic and geographical sense of the term. We will use the example of bocage grasslands to illustrate the complexity of landscape dynamics. Secondly, in order to identify more precisely which forms of heritage are benefiting from restoration projects today, we will study a concrete case by attempting to retrace the geo-historical trajectory of different areas that have now been "restored" on a territory scale (a portion of the Val de Loire). Finally, in view of the complexity observed, we will re-examine the meaning of baselines for agro-pastoral landscapes: should they really be considered as historical states or rather as very contemporary forms, testifying to a new relationship with nature and constituting new baselines for development?

11.2. How does an agrarian landscape evolve?

11.2.1. *Nested subsystems*

In order to understand the dynamics of an agricultural landscape, it must first be considered as a system comprising several interlocking sub-systems. In this complex assembly, there are, on the one hand, the objects, *i.e.*, the landscape structures and

3 The separation inherited from the Romans between the *hortus* (garden), the *ager* (cultivated fields), the *saltus* (grazed areas) and the *sylva* (forest) corresponded from the start to a gradient of artificiality.

4 2015 figures given by the French Ministry of the Environment.

forms that make up the agricultural space and, on the other, the views that are brought to bear on these objects, *i.e.*, the values, expectations and esthetic judgments specific to an era and a society. This separation is useful for the study, but it remains artificial, since the way in which the objects are read already depends in part on the way in which they are viewed[5].

In order to understand landscape forms and structures, we will rely on rural geography, which made agrarian landscapes its main object of study during the greater part of the 20th century. Inspired by Meynier, Juilliard, Dion, Panhol and all of his predecessors, René Lebeau established the basis for this reading in "*structures agraires dans le monde*" (Lebeau 1969), first published in 1969 and republished many times since. It is from this reference work that we draw the following definitions.

An agrarian landscape can be broken down into four subsystems:

– first, the "**terroir**", in the geographical sense of the term, *i.e.*, all the physical conditions (relief, water, soil, climate, exposure, etc.) that make up an agronomic unit, which may already have undergone numerous changes (drainage, irrigation, development of terraces, etc.). The terroir is the foundation of the production systems and, generally, several terroirs are combined within the same farm;

– second, the "**rural habitat**" which, through its forms and layout, reflects multiple strategies of adaptation, exploitation, protection, etc., at different scales. On a communal scale, the habitat can be grouped, dispersed or gathered in hamlets according to regional particularities. On the scale of a farm, the architecture of the dwelling house and annexes, their organization in space and the materials used are once again indicators that inform us about the logic of production, the nature of the subsoil, the climate, the economy or local beliefs;

– third, to be used, the land is divided into blocks or plots. Roads and paths are laid out to connect the inhabited areas to each other and to the plots. This division of space constitutes the "**agrarian morphology**" and can vary enormously from one region to another: the plots can be more or less geometric, with shapes that follow the relief of the land or not, and of very different sizes, depending on the ownership of the land, the demographic pressure and the care given to each crop;

– fourth, the "**cropping system**" is the finishing touch, adapting to the soil and climatic conditions, to the skeleton formed by the division of the fields, to the constraints of operation and to the highly fluctuating economic market. In this

5 See the European Landscape Convention.

geographical definition, the cropping system is not only the range of crops grown and their rotation (the crop rotation), but it also includes the tools, techniques and know-how directly related to these crops.

Ideally, we could imagine a landscape system that functions in a stable manner and totally in accordance with local conditions: the cultivated plants being perfectly adapted to the properties of each terroir, the farm buildings designed according to the bio-climatic conditions to shelter all the harvests as well as the production and transformation tools, and then the plot division optimized according to the qualities of the soils and the means of use. Today, we are certainly nostalgic for this balance when we regret the "traditional landscapes" but has the perfect blend ever existed?

11.2.2. *A system in constant evolution*

Like any dynamic system subject to both internal changes and external influences (new production standards and new consumer expectations, for example), agrarian landscapes are in perpetual evolution with very different rhythms for each component. Of the four subsystems mentioned, the **terroirs** seem to form the most stable part, subject only to climatic variations and to phenomena of orogenesis and pedogenesis when they are not intentionally modified. However, over time, certain developments have been able to transform them more brutally, for example, when marshes were drained. The **habitat** also has a certain stability, but it can undergo spectacular changes in response to an expansion of farms, to new construction standards or after episodes of destruction[6]. Since **agrarian morphology** is a direct reflection of land ownership, it varies more easily, especially at each generation change. For example, over the course of history, we know of major cadastral redistribution operations that have completely disrupted landscapes: the first reparcelling after the Hundred Years' War, the division of communal land in France during the 19th century or the reparcelling of the 1960s, for example (Figure 11.1). However, as archeogeography has shown, medieval parceling often continues to influence current land divisions (Watteaux 2012), according to a process of "transformission" (Chouquer 2007). Finally, the **cultivation system** is the most changing part of the system, on the one hand, because the evolution of techniques more easily follows the modernization of society and, on the other hand, because the choice of productions is very dependent on the economy and consumer expectations.

6 For example, the flood of 1856 caused the disappearance of the pisé houses that occupied the Authion valley.

Thus, over time, several cropping systems have succeeded one another on the same landscape skeleton. However, there comes a time when adaptation is no longer possible and the evolution of cropping systems then generates transformations of other parts of the system.

Basically, agrarian landscapes do not evolve in a linear fashion. In order to account for these complex trajectories, it is necessary to establish some fundamental principles, borrowed from geohistory. This discipline relies on both spatial and archival data to attempt to trace the evolution of a space. It is not limited to a "geography of events" (Elissalde 2000) that would attempt to place all the stages of construction of a landscape on a single line, as if the physiognomy of the latter resulted solely from programmed actions and a succession of moments of balance and division. On the contrary, it recognizes a multitude of temporalities that collide and of which the landscape forms are only imperfect mirrors at a given moment. There is therefore always an element of uncertainty in the rewriting of trajectories that must be accepted (Gramond 2014). Studies on the long term have particularly demonstrated this difficulty: it is the complex interactions between facts of very different orders that create forms (Chouquer 2007), and it is not uncommon for apparent stability to hide underlying dynamics which are already preparing future transformations. This phenomenon leads to periods of latency that can be explained by the concept of resilience, *i.e.*, the capacity of landscape forms and structures to absorb a disturbance without apparent modification.

To illustrate this complex trajectory of agrarian landscapes, we will take the example of a legacy that is very present in the west of France, particularly scrutinized by landscape ecology and often the subject of restoration policies: the so-called "natural" grasslands surrounded by hedgerows.

11.3. Example of the natural grassland–hedgerows combination

In order to understand the origins of this combination of natural grasslands and hedgerows, it is necessary to dissociate the evolution of parcel divisions, of which hedgerows form the boundaries, from the evolution of forage areas, of which grasslands are a form of production. We can try to schematize these two intertwined histories (Figures 11.1 and 11.2).

11.3.1. *Initially, a few hedge fences around the cultivated fields*

At the end of the Middle Ages, individual grasslands surrounded by hedges were not very common, except in the immediate vicinity of the houses, in order to watch over the farm animals and pen the working animals. The grazing of the herds was mainly done away from the cultivated land in dedicated spaces (*saltus* reclaimed from the forest) or in alternation with ploughing, thanks to the introduction of free grazing on fallow land[7] or after the harvest. The moors, "courses", "marshes" and "pastures" that made up the permanent pastures represented a heterogeneous group of poorly maintained spaces, the use of which was most often collective in two main forms: either they were community pastures (called "common", "communal", etc.), or they were grasslands that were mowed individually as well as subject to the "common pasture", *i.e.*, they were open to all for the second grass. This community regime guaranteed both access to the pastures for as many people as possible and their fertilization by as many animals as possible but, as a result, prohibited fences. Conversely, fences protected cultivated fields. However, the small size of the plots of land suggests that they were collective hedges surrounding islands of open fields, forming small open fields managed in community near the hamlets: "gagneries" (Charaud 1949; Glomot 2011). The whole of this organization had to serve one and the same purpose: the production of cereals.

11.3.2. *Division of the soil and multiplication of fences*

From the 16th century onwards, a first consolidation of the large seigneurial domains and the retreat of the woods led to the installation of new independent farmsteads of a larger size than the old houses in enclosed grounds. They seemed to be behind the first proliferation of fences, on the borders of the moors and away from the first tenures. This movement, initiated by the nobility and the clergy, followed that which took place in England between the 15th and 18th centuries (Poussou 1999), and which consisted of sharing communal lands to increase forage production; all with the aim of having more animals and increasing the quantity of manure, which was indispensable for boosting grain production. This is how the "enclosure" movement was born, which led to a progressive partitioning of the commons throughout the United Kingdom by means of dry stone walls, ditches or hedges[8]. This enclosure was therefore a first factor of progress, leading to the intensification of cereal production while feeding a textile proto-industry from sheep wool.

7 The fallow land, strictly speaking, designated a resting land but worked in preparation of a next sowing and previously opened to grazing (see the works of François Sigault).

8 The Inclosure Acts were the laws that progressively imposed the enclosure of fields in England between 1760 and 1840.

In France, it was not until the Revolution and the confiscation of seigneurial property that the same movement could be initiated. From the first half of the 19th century, many communal lands (pastures, moors and uncultivated lands) were thus divided and sold to the wealthiest farmers, who hastened to install fences. As the divisions were organized collectively by the new municipalities, they gave rise to a very geometric form of division: the "mimetic" bocage, as opposed to the old forms, which were more "organic", *i.e.*, following the topography (Palierne 1976). The abolition of grazing (officially authorized by the law of 1890) also promoted the partitioning of the land.

11.3.3. *Extension of fodder crops*

In spite of everything, in the heart of the bocage countries, in the first half of the 19th century, the majority of the hedges continued to surround the crops, always grouped in islands when the divisions were too significant. The enclosed plots were then often surrounded by a grassy strip: a "*perimeter of fields*" (locally called "*cheintre*", "*chancière*", "*tournière*", etc.), sometimes planted with fruit trees, which allowed the ploughs to turn but which also had the function of protecting the crops from the undesirable effects of the plant hedges. The replacement of annual crops by permanent grasslands took place in two stages.

The first revolution, which began in England in the 18th century, consisted of introducing fodder plants into crop rotation to replace fallow land: so-called "artificial" grasslands[9], as opposed to permanent or temporary[10] "natural (growing) grasslands". This real revolution was again aimed at increasing cereal production by better feeding the herds. In France, statistical surveys (Block 1860; Ministère de l'Agriculture 1892/1936; Agreste) indicate a spectacular expansion of artificial grasslands between 1840 and 1892, when they represented 6.25% and then 12.5% of plowed land, respectively. The surface area then remained at around 3.2 million hectares until 1950 (*i.e.*, 16.5% of the land still plowed) (Figure 11.2.).

9 Consisting of alfalfa, sainfoin, clover, ryegrass, etc.

10 A temporary grassland is sown with grasses but turned over before five years.

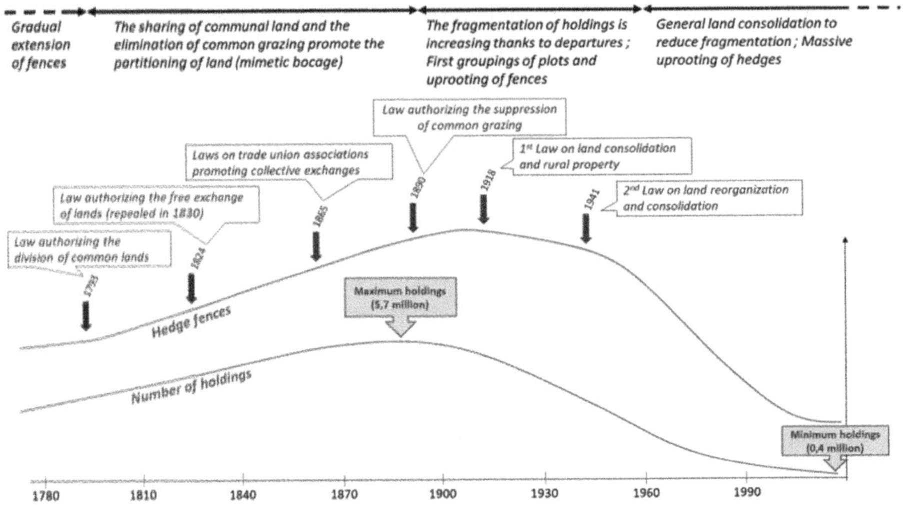

Figure 11.1. *Trends in the evolution of hedgerow enclosures and land fragmentation since 1789. For a color version of this figure, see www.iste.co.uk/godet/baselines.zip*

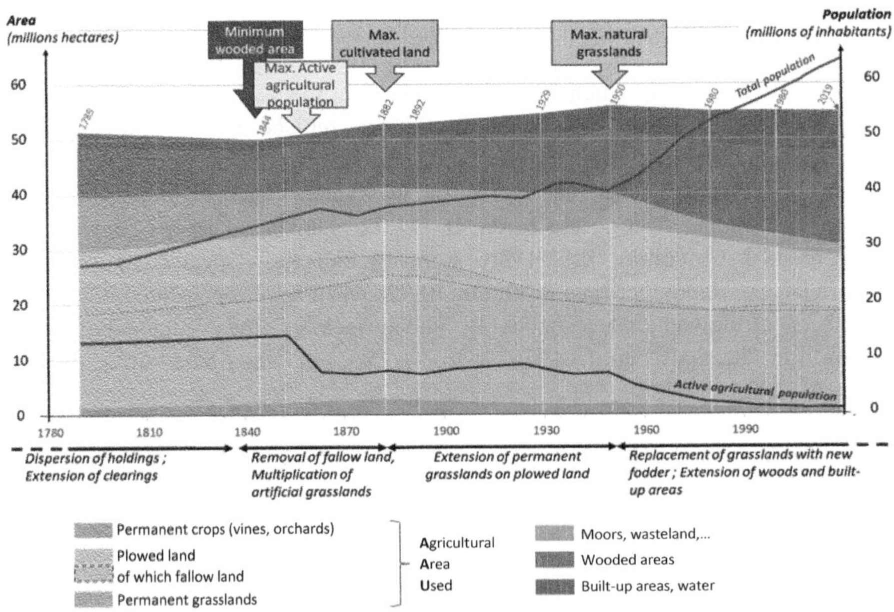

Figure 11.2. *Evolution of land use and population in metropolitan France between 1789 and 2019. For a color version of this figure, see www.iste.co.uk/godet/baselines.zip*

In a second phase, in connection with the Industrial Revolution, this forage transformation took another turn, aiming at a second objective. It was no longer just a matter of increasing the production of cereals, which could now benefit from other improvements, and of providing meat and dairy products to feed the growing cities. A number of factors encouraged the development of livestock production in the western regions (Hédin 1951). As a result, many plots of land were converted to grassland. Statistics allow us to account for the extent of the change. In metropolitan France, for example, areas still under grass more than tripled between 1840 and 1950, while arable land decreased by a quarter. In Normandy, this has also resulted in the disappearance of thatched roofs...

11.3.4. *The hedge fence loses its usefulness, then becomes a nuisance*

However, by the end of the 19th century, as pastures (temporary and permanent) progressed, to the detriment of cereal crops, hedge fences became less useful. The partitioning of the pastures – always encouraged by agronomists – could henceforth be satisfied with a simple wire imitating the bramble[11]. Moreover, the more livestock farming developed, the more valuable the land became, and the more space lost to hedgerows, embankments, ditches and sunken roads represented a significant loss for farmers (Poirier 1934). However, above all, the division of the land that had accompanied the multiplication of farms and the access of farmers to property was coupled with an increasingly troublesome parcel fragmentation. The latter, initially encouraged by the equitable sharing of inheritances, the need for crop rotation and the need to distribute crops among the different terroirs, was now amplified by the enlargement of farms integrating tiny scattered plots, freed by those who left the land. The laws of 1865 and 1888 on syndicated associations to promote agricultural improvements should have led to exchanges of plots, but they had little success (Schmerber 1950). Some hedgerows began to be torn out around 1900 (Chevalier 1949). Then, a first law on the consolidation of rural properties, called the Chauveau law, was voted on in the Senate in 1918. However, while the bocage structures did not really regress until the 1960s, it was certainly because the hedges – and no longer the fences – had acquired new functions. In particular, the introduction of animals in the heart of the hedgerows had revealed the advantage of their shelter. Increasingly wooded, they also represented a significant resource of wood for heating and construction (Chevalier 1949), all the more indispensable in regions of small-scale cultivation, where forests had declined[12] enormously. But when this resource was itself devalued by the use of new energies (coal, fuel, electricity) and

11 Barbed wire was invented in 1874 by an Illinois farmer.

12 For example, forest covered only 3.2% of the Manche department and 6.8% of the Morbihan department in 1950, compared to 18% for the whole of France.

the use of new construction materials (iron, concrete), nothing seemed to oppose their removal.

Finally, the rural exodus had also rendered the farming methods inherited from previous generations obsolete. Between 1850 and 1950, mechanization – at first, to compensate for the lack of manpower – the development of the fertilizer industry, the improvement of seeds and animal breeds were all techniques that allowed farmers to adapt. After World War II, this modernization became an additional argument for redistributing land. The large-scale reparcelling operations then lead to a second fodder revolution, this time consisting of turning over permanent grasslands (Chazal and Dumont 1954).

11.3.5. *In the end, which baseline(s) should be retained?*

This complex example shows us that agrarian landscapes did not evolve in a linear fashion, and although there were significant episodes, they cannot be considered alone as turning points that would have marked the end of a balance. Thus, the major reorganizations of the 1960s and 1970s are not simply a reflection of post-war agricultural policies, but must be understood as an almost inevitable consequence of the growing gap that had developed between the agrarian morphology inherited from the 19th century and the modernization of farming systems. These land restructurings, often accompanied by operations to "improve" the land, were therefore simply a rebalancing of the system in a continual search for adaptation and progress. In these conditions, it becomes difficult to isolate certain events or certain dynamics and hold them responsible for the degradation of a landscape. For example, should we denounce the Industrial Revolution, when it can be considered as one of the essential reasons for the multiplication of grasslands for more than a century? Should we regret the intensification of agriculture when it is with this same objective that many hedgerow fences were established until the end of the 19th century? Should we still condemn livestock farming, which is the only way to maintain grasslands? Each landscape form is a hybrid form of heritage and to us it seems very complicated to isolate in a historical trajectory a state of the system that could constitute a baseline to be reproduced.

11.4. What exactly do the restored spaces represent?

While there is no simple past for the bocage grassland, we can then ask ourselves what baselines guide restoration projects today. In concrete terms, what are the agrarian legacies that we are trying to conserve or restore? What do they correspond to?

11.4.1. *A mosaic of spaces*

To answer these questions, we will once again start from a concrete case by taking the example of a territory that is both highly developed and well documented, in which we can attempt – with all due caution – to trace the trajectory of protected areas. The area chosen is a portion of the Loire Valley (the Authion Valley), which has been listed as a UNESCO World Heritage Site since 2000 as a "cultural landscape", but is also increasingly valued as a form of "natural heritage". The Val d'Authion is located in the Maine-et-Loire, at the gates of Angers. Its downstream part, called the "Brain marsh", remained in a state of flooded grasslands until the 1970s because of both the frequent overflows of the Loire and the backflow of the river into the Authion, its tributary. These wet grasslands, generally perceived as a homogeneous whole, represent, however, a multitude of complex legacies. On the 1949 photo, several of them can be distinguished (Figure 11.3).

1 Grasslands excessively divided and subject to common grazing until 1970
2 Grasslands/lands made "advantageous" and fenced off before 1800
3 "Grassland marshes" reclaimed from the communes, drained before 1800
4 Communes divided around 1835 then fenced off by hedges
5 Private willow plantations separated by hedges and ditches
6 Lands annexed to the communes in the 19th century

Figure 11.3. *A great diversity of grassland legacies in the Val d'Authion in 1949. For a color version of this figure, see www.iste.co.uk/godet/baselines.zip*

This varied group then underwent a vast development project during the 1970s, led by the former Minister of Agriculture and Maine-et-Loire deputy Edgar Pisani, aiming to "clean up the marsh" in order to create a horticultural center. The territory was thus radically transformed, reorganized, drained, etc. However, in spite of all this, there are still some grassland legacies. We will study three areas that are now benefiting from ecological restoration: the "Andard reed bed", the "marsh of the upstream grasslands" and the "Rillé lake", further upstream. The comparison of their landscape trajectories allows us to provide some answers to the questions posed (Figure 11.4).

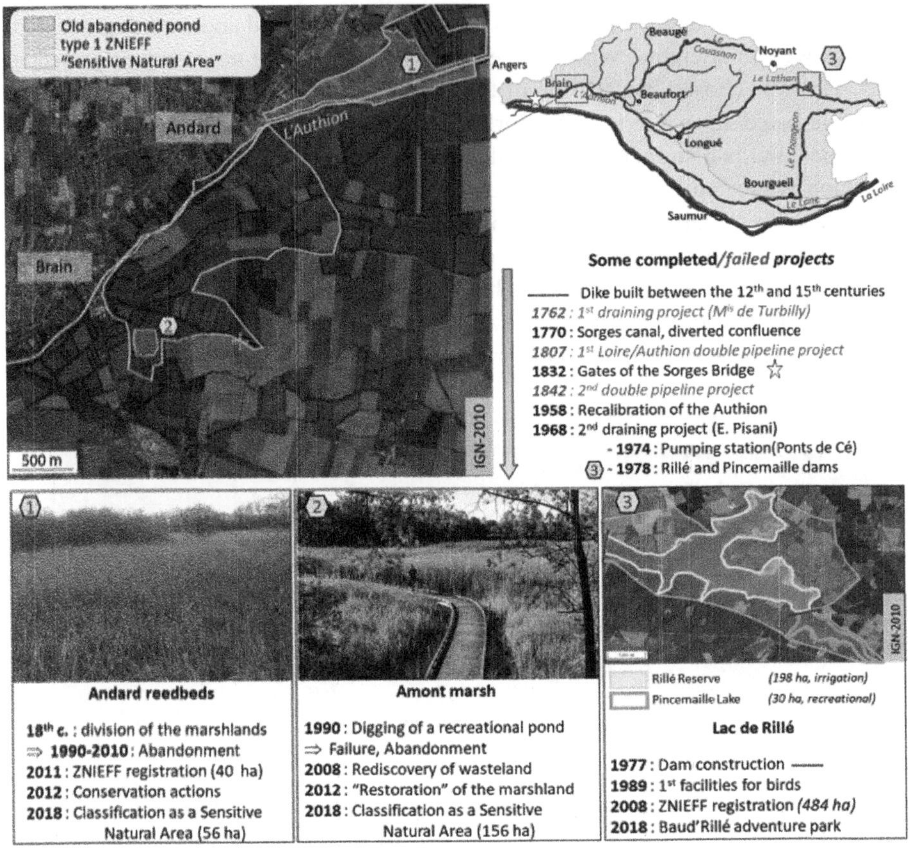

Figure 11.4. *Draining and ecological restoration projects in the Authion basin. For a color version of this figure, see www.iste.co.uk/godet/baselines.zip*

11.4.2. *Inability to identify a historical baseline*

First, it is not possible to identify a single baseline in this territory that would justify the choice of spaces to be conserved or restored. On the contrary, each of them seems to have been inherited from very different periods of development according to temporalities and spatial scales that are themselves excessively varied. What do the Andard reedbeds, inherited from the progressive abandonment of wet grasslands surrounded by ditches during the 18th century, have in common with the reedbeds of the "Amont marsh", resulting from the abandonment of a pond dug in the 1990s? What is the similarity between this abandoned pond, the result of a local enterprise, and the Rillé lake, inherited from a vast national land development operation around 1970?

Second, even if it seems possible a priori to date each of these landscape legacies by taking the development operation that led to their establishment as the origin, the current state does not correspond to this initial state and the forms restored today are not the forms that were established. Each one is the result of its own dynamics, which also testify to phases of abandonment, which are marked to varying degrees. Moreover, each development operation that seems to have initiated these forms is itself inherited from older logics, which again are difficult to date. For example, Edgard Pisani's project, which led to the digging of the Rillé lake, the cultivation of the Brain marsh and, indirectly, the abandonment of the Andard grasslands – by reducing the number of livestock – is itself to be linked to the multiple attempts at drainage that followed one another over the course of the previous centuries.

Therefore, even if there were significant events in the history of this territory, it is again difficult to identify a stage which would be completely "deviating". The draining of the Brain marsh in the 1970s has certainly led to a clear retreat of the grasslands but it is also thanks to this project that upstream what is considered today as the biggest ornithological reserve of the Center region has been created. This fact is all the more remarkable given that, originally, the lake had been conceived in two parts: upstream, a first basin of 30 hectares intended for nature recreation to maintain a full tourist activity and, downstream, the reserve of 198 hectares, only intended to regulate the waters in the Authion basin and supposed to be less rich and less attractive, because of the seasonal variations.

Finally, in these three examples, the baselines are above all current states, more or less random legacies of a combination of agrarian developments dating from very varied periods, phases of partial abandonment according to equally varied temporalities, and a tourist dynamic around the wetlands initiated in the mid-1960s, in connection with peri-urbanization (Bonneau *et al.* 1974). However, these baseline landscapes are also symbolic constructions which are also very contemporary. Each of these protected forms is, in fact, the bearer of a narrative that makes it correspond more or less perfectly to the expectations of the moment: it is in "good ecological health", rich in threatened or emblematic species, accessible to the population and more or less staged to correspond to the ideal image sought. The project of the "renaissance of the Amont marsh", for example, tends to make this place correspond to a marsh model covered with reeds such as the population could imagine it. However, despite its name, the place was never this marsh, neither before its deforestation, nor after its transformation into pastures.

11.4.3. *So what exactly do these baselines represent?*

How can we name all these complex legacies? The vocabulary commonly used is often imprecise, oscillating between highlighting their natural dimension ("natural sites", "natural environments", "sensitive natural areas") and a more neutral name to designate their new status within agricultural areas ("areas of ecological interest", "topographical elements", according to the PAC). The term "semi-natural" dominates in planning documents to designate this in-between area.

Ecologically, each agro-pastoral legacy constitutes an anthropogenic ecosystem, in the sense defined by Tansley in 1935, *i.e.*, a "unit of nature on the face of the earth" composed of living organisms (a biocenosis) and inorganic elements (a biotope) in permanent interaction, but whose evolution has been strongly influenced by human activities. In other words, they are microsystems in apparent stability, resulting from multiple allogenic successions (corresponding to agronomic events) or autogenic successions (when there has been abandonment), which are regressive or progressive (depending on the manager's point of view). These are therefore "quasi-equilibria" whose only common feature is an interesting, even remarkable or heritage biodiversity.

However, from an agronomic point of view, these are only extracts, or even fragments of landscapes: more or less random states resulting from a combination of a terroir, an agrarian morphology and cropping systems at a given moment in their respective trajectories. These are ephemeral states, fragments of history that are particularly difficult to link to an agricultural memory for several reasons. First, because, as we have seen, these fragments are under the permanent influence of varied and multi-scale dynamics that do not allow them to be associated with a clearly defined historical period. Second, because they have acquired new ecological values, these landscape forms are interpreted according to a grammar and vocabulary that make their agricultural past difficult to read. For example, the term "natural grasslands" today includes, without distinction, mowed grasslands and grazed grasslands, the "courses" of the plateaus of the Pays de Caux and the "*herbus*" of the banks of the Seine, the "*pâtis*" of the valley bottoms or the "*cotis*" of the seashores, etc. (Hédin 1951). In short, all the old classifications linked to the use of these grasslands disappear behind a single term designating a form of ecological habitat – the meaning of the adjective "natural" having itself changed. Similarly, the new functions attributed to hedgerows (refuge and corridor effects in particular) contribute to reinventing typologies that completely erase the complex history of these objects. For example, the anti-mistral barriers introduced along the Rhone valley during the 20th century and composed exclusively of cypress or white poplar trees are interpreted in the same way as the fences on the slopes of the western bocages. The former are then reproached for their singularities (mono-specific composition, absence of "connectivity") which are judged as degraded[13] states. In the same way, we have shown how the history of the pollarded ash bocage of the Loire Valley had been completely rewritten by projecting contemporary expectations of nature onto forms inherited from the past (Montembault 2007b). Finally, what can be said of the term "green corridors", which indiscriminately mixes agrarian structures with totally disparate origins and functions and which, at the same time, excludes the entire panoply of annual crops. This ecological rereading thus tends to dismember agrarian landscapes by making selections and denying the mosaic of cultures, which becomes perceived as a "threatening fragmentation" (Agnoletti 2014).

11.5. What to do then with this forgotten or reinvented past?

11.5.1. *Three possibilities*

By reinterpreting inherited agrarian forms, renaturation operations thus tend more often than not to distort agricultural memory – or even to deny the cultural value of these legacies altogether (Antrop 2005; Agnoletti, 2014). What status, then, should truly be accorded to "baselines" in renaturation operations? Three avenues seem conceivable.

First, maintaining the opposition between natural and anthropic. However, this is equivalent to considering the baseline as an original state, undisturbed by human activities; this does not make sense for an agrarian landscape. This concept seems applicable only to rewilding and "passive restoration" operations that aim to erase all forms of anthropization.

Second, accepting, on the other hand, that going beyond this first opposition allows the baseline to be assimilated to a stage of evolution in the trajectory of the landscapes, *i.e.*, a state of disturbance that is deemed acceptable, before "degradation". This is the position that seems to be unanimously accepted by managers, and also defended by many authors. However, all recognize the difficulty of defining this historical baseline: "how far back in time should we go: a few years, decades, centuries, millennia?" (Bioret 2011). But this is not the option that we defend, as it seems illusory to freeze a landscape system. From an ecological point of view, isolating an ecosystem (in this case, an agrarian structure) is also an artificial process that certainly allows science to make it an object of study, but does not guarantee its conservation. Tansley already underlined this limitation in 1935,

13 See PNR Alpilles (2015).

when he wrote: "the systems we isolate mentally are not only included as parts of larger ones, but they also overlap, interlock and interact with one another [...] Relatively to the more stable systems, the ecosystems are extremely vulnerable, both on account of their own unstable components and because they are very liable to invasion by the components of other systems". While it seems futile to us to try to conserve or recover a past state, there is, however, another way to consider the baseline.

A third alternative consists of no longer seeing the baseline as a historical landmark, but rather as a spatial landmark, a kind of control environment for (re)naturation policies. This change of paradigm then obliges us to assume the modern character of the preserved landscapes, as much in their forms as in the functions and values that are attributed to them. In spite of everything, the question of the remobilization of legacies remains intact and two strategies can once again be opposed: either the "witness environments" are extracted from an agricultural logic and it is, in this case, the conservation of the ecological memory that is favored, or, on the other hand, they are integrated into new productive spaces, by prolonging an agronomic memory of the places. The two approaches do not produce the same effects.

11.5.2. *Maintaining an ecological memory at all costs*

The ecological[14] memory of a place can be understood in two forms: first as a legacy as well as the driving force of an ecosystem in perpetual evolution[15]. The desire to preserve a state of equilibrium inherited, in part, from human activities means assuming the need to intervene – sometimes in a spectacular way – to promote certain species and/or eradicate others that are deemed more or less "invasive". These are therefore "non-neutral selective processes" (Gramond 2014) that are particularly demanding when it comes to maintaining fragile balances. However, it turns out that it is often these fragile and transient states that are the focus of all attention – both because they are found on the edge of productive spaces and because they constitute rich and/or rare pioneer habitats. Regardless of cost, this form of controlled restoration is therefore widespread. Balaguer *et al.* (2014) thus propose to draw on the past to "guide the ecosystem in a consciously chosen pattern", recognizing that "after all, ecological restoration interventions are simply another round of human-caused disturbance, but in a new sense and spirit". This

14 The concept of Ecological Memory (EM), from medicine, has recently been applied to restoration ecology (see Sun and Ren 2011).

15 For example, the flora associated with a prairie is both a testimony to the past and an essential element of future transformations, through the seed bank stored in the soil.

selection operation is therefore very similar to an agricultural activity whose aim is no longer to promote edible, medicinal or dye species but rare and/or heritage species. Farmers are often asked – as in the Val d'Authion – to maintain the protected areas at regular intervals by mowing or "eco-pasturing". In other cases, the agro-pastoral legacy remains in a production circuit, but since priority is given to the conservation of an emblematic species or the maintenance of a particular biocenosis, the maintenance rules imposed on farmers are increasingly strict and progressively disconnected from agricultural requirements. For example, within the framework of agri-environmental contracts set up in the valleys of the Anjou region, the reduction of mowing in order to encourage the nesting of corn crakes at fixed and increasingly late dates, regardless of the weather conditions, has contributed, in some years, to a significant drop in the nutritional value of the hay harvested (Montembault 2007a). Similarly, in the Pyrenees, the imposition of new rules that are compatible with the introduction of the bear or the development of tourism makes the work of breeders increasingly difficult (Eychenne 2018).

This mode of conservation that aims at maintaining a predefined ecological balance therefore has a high cost, since it is less and less compensated by an agri-food production. In itself, this is not a problem, if society is aware of it and if it is ready to assume that it is indeed a form of gardening (Collomb 1977) that aims to build new spaces with both ecological and recreational functions through a skillful control of living things. But presenting these projects as a return to a more natural past suggests a form of thrifty, even autonomous, management (wouldn't the grassland grow naturally?), which distorts the debate. Moreover, another problem arises when we reason on a larger scale. Removing land from agriculture encourages the intensification of production outside protected areas and even outside Europe, without necessarily controlling the ecological impact of these relocations. Between 2000 and 2010, for example, Europe doubled its imports in order to reach the equivalent of what 35% of virtual agricultural land would produce (Von Witzke and Noleppa *et al.* 2010). Can we not then imagine other forms of renaturation that better integrate agrarian legacies?

11.5.3. *Extending the agricultural memory and accepting the creation of new agro-pastoral landscapes*

Preserving the memory of a landscape does not necessarily mean preserving its forms, which, as we have seen, are constantly changing and being reinterpreted. It is these complex dynamics that can explain the great variety of agrarian landscapes that we have inherited. Let us recall that, in 1946, France was divided into 432 agricultural regions (Klatzmann 1955), largely inherited from the small geographic regions, themselves derived from the Roman *pagus*. Behind this division, which is still used today, we can guess the rich mosaic of landscape assemblies – and even more so, that of ecosystems – that reflected each small "region" of the past. The originality of traditional agro-pastoral landscapes is therefore not limited to particular structures any more than traditional rural societies can be associated with a single mode of organization (Mendras and Jollivet 1965). On the contrary, regional diversity was one of the main characteristics of the French and European countryside at the beginning of the 19th century. However, this variety of landscape models does not mean that there were not commonalities between all the underlying socio-ecological systems[16]. It is therefore by questioning the evolution of these systems and, more specifically, the influence of natural factors on the development of rural societies and the planning of spaces that we can consider other ways of extending the memory of traditional agro-pastoral landscapes.

In 1975, Georges Bertrand was already advocating an ecological history of rural France, reminding us that nature should not be studied separately but "as a lived reality, both dominant and dominated, fought and used from within a social and economic organization" (Bertrand 1975). What then are the elements that could characterize the links between so-called "traditional" rural societies and nature? From when or at what agro-technical threshold can we consider that they have entered a form of modernity? Once again, it seems illusory to look for a single point of division, even if Enlightenment thinking and the gradual desacralization of nature are sometimes presented as accelerators (Hurand 2014) of a process begun at the origins of agriculture and consisting of making humans increasingly the master of their environment. However, by questioning the functioning of rural societies at their peak – *i.e.*, for France, between 1830 and 1850 – we can attempt to identify what strongly linked them to their environment and which it might be interesting to recover. The exercise is perilous because it is not a question of idealizing the past nor of projecting a very positive contemporary vision of nature, forgetting all the negative aspects that pushed the traditional populations to want to extract themselves from it. However, at the risk of caricature, we can nevertheless retain some fundamental traits that underline the difference between past and present landscape dynamics.

The first element that emerges from the interpretation of ancient agrarian landscapes is the influence of local human and natural factors in the adjustment of production systems. It is partly to this predominance of the local over the global that we owe the great diversity of rural landscapes. The influence of the environment – without it having been a determining factor – can be seen in all the components of

16 See Davasse (2015).

the ancient landscape systems: in the layout of the rural habitat, in its architecture and construction materials, in the division of fields and the distribution of roads, in crop rotations, etc. This emphasis on natural factors reflected, above all, a need for autonomy, which was particularly important in landlocked territories, located away from cities and major river or maritime communication routes. Under these constraints, the farming communities had to show creativity in order to use all the resources at their disposal by adopting techniques that could be described before the time as "agroecological engineering". The multiplication of hedgerows, *i.e.*, the construction of fences from so-called "indigenous" plants, to increase the number of livestock and thus the production of fertilizer, is a good illustration. But this dependence on local resources also required forms of regulation at all scales in order to limit overexploitation. This role was ensured by rural leases, collective exploitation rules, ancient customary law and certain royal directives, such as the 1669 ordinance written by Colbert, aimed at protecting and restoring the wood[17] resource. Finally, let us not forget that, until the Revolution, the systematic taxation of harvests encouraged farmers to produce little – and therefore to live in extreme sobriety (Blanqui 1851).

Second, what is also striking in the old rural landscapes is the very strong articulation between the scales and between all the landscape components. This great cohesion of systems testifies to a nature that was once daily and entirely inhabited, shaped on a human scale, where production spaces were the extension of living spaces and where cattle and farmers lived side by side, sometimes in the same house. This embodiment of places can be seen, in particular, in the old units of measurement which, from antiquity until the introduction of the metric system (in 1795 in France), were based both on the dimensions of the human body[18] and on its ability to act in its immediate environment[19]. In a way, it was the capacity of interaction between a human being and their environment that governed the organization of spaces and their sizing. We can then guess at the local and evolutionary character of these units, which did not facilitate the exchanges between the regions. In the same way, the measurement of time and durations rested entirely on the observation of the natural cycles, and this has been, since antiquity: the rotation of the stars, of course, as well as cycles of life and variation of the environments. In ancient Egypt, for example, a year corresponded to the time necessary for a harvest and included only three seasons: the Nile floods ("Akhet"), the flood and sowing ("Peret") and the harvest ("Shemou"). Among the Romans, the

17 Ordinance of Louis IV "on the fact of waters and forests" (August 1669).

18 For example, the inch, the foot, the *toise* and their different multiples: the perch, the arpent, etc.

19 A "league" represented the distance covered on foot in an hour or half an hour, a "journal" or an "acre", the area ploughed in a day, a "*boisselée*", the area sown by a bushel of grain, itself measured by a container that could be transported by men's arms, etc.

year began at the same time as the rebirth of nature, *i.e.*, in March and, in our Gregorian calendar, inherited from the Julian calendar, the pagan festivals as well as the religious festivals were largely aligned with the ancient rhythm of sowing and harvesting. Finally, before the advent of the railroad, which made the adoption of legal hours at the scale of nations necessary, then the establishment of a universal time system (adopted in France in 1911), the only time used in the countryside was the one given by the sundials (the true solar time), only simplified during the 19th century into average[20] local times. Until the dawn of the 20th century and the erasure of local particularities in a movement of affirmation of national unity (Weber 1986), the work of the fields, as well as all of rural life, took place at the rhythms of nature, a nature experienced from the inside, daily and meticulously scrutinized to draw the maximum of advantages from it while protecting it from its excess. It is therefore, above all, this mode of operation that the ancient rural landscapes tell us about, and if we had to retain a few fundamental traits, they would be the complementarity between a great diversity of spaces shaped by human hands according to rhythms and configurations adapted to each terroir (without recourse to chemistry), the extension of productive spaces to all territories according to gradients of artificialization, going from the most meticulously cultivated lands (gardens, orchards) to more extensive forms of exploitation (pastures, moors, woods), the optimal use of local resources and materials – with particular importance given to the trees in the fields – or the absence of a clear border between living spaces, production areas and places of relaxation, all of which evolve together, slowly but continuously.

The agrarian structures of the past thus reveal ways of living and using nature that are very different from what we know today. While there is no question of returning to the discomfort of these lifestyles, which were overly dependent on natural hazards (the average life span in France was 25 years around 1750 and only 43 years in 1850[21]), it would nevertheless be interesting to reconsider the genesis of landscape legacies in order to think of more complete ecological restoration processes. Thus, rather than seeking to maintain certain landscape structures at all costs – to the point of making them "natural spaces" that are isolated from everything – could we not incorporate production systems that respect the environment and natural rhythms by allowing ecosystems to evolve in a more random manner? This is, to some extent, what has already been implemented through certain innovative approaches aimed at making the best use of the potential of the environment by adapting crop systems to each situation and limiting external inputs. This is clearly the initial spirit of permaculture, for example, or of

20 In 1826, Paris adopted the "local mean time", which then became the time of the railroads until France joined the world time system based on Greenwich in 1911.

21 Figures from the *Institut National d'Etudes Démographiques*.

eco-growing, even if these systems do not refer directly to the past and advantageously integrate modern knowledge and technologies. Agro-forestry is another extension of ancient processes consisting of combining various productions to take advantage of a set of resources (water, soil, sunshine, etc.) without exhausting a land. In all cases, new landscape forms emerge, new ecosystems are created and new dynamics are put in place, going far beyond the logic of greening.

11.6. Conclusion

Under the eye of the geographer, landscapes are only imperfect reflections of deeper dynamics. What should therefore be preserved of traditional agrarian landscapes? The great decline of rural societies that began in the West in the first half of the 19th century led in turn to the end of "regions" from 1870 onwards (Weber 1983), then to the end of peasants (Mendras 1967), and finally to the death of traditional agro-pastoral landscapes (Dagognet 1982; FNSAFER 2004). This radical transformation of Western societies was accompanied by a form of disconnection from the surrounding nature and has also manifested itself in a general impoverishment of biodiversity. In view of the most recent landscape transformations – which are also the most dramatic – it is tempting to want to save at all costs certain objects or certain assemblies inherited from the past, such as ponds, bocage grasslands, moors, etc. However, a close look at this more or less distant past shows that the ecosystems that we preserve today are often only transitory states, modern configurations resulting from phases of development and temporary abandonment that have followed one another in a complex and random manner. The maintenance of these fragile balances is therefore a device, a new form of gardening whose main goal is no longer to produce "useful" plants but to conserve a certain type of biodiversity. While these actions may seem legitimate in the sense of preserving the ecological memory of certain territories and/or protecting a few emblematic species, they are, on the other hand, very costly, and all the more so when they are disconnected from agro-forestry production. The protection of remarkable spaces also often goes hand in hand with the development of recreational activities, known as "nature", and the installation of equipment to guide walking and discovery. These are all places that have been transformed for leisure purposes and removed from agricultural space, according to very contemporary processes that, in turn, reflect the profound mutation of rural spaces. For the majority of city dwellers, the countryside has become a recreational space (Hervieu *et al* 1996), a place where it is good to live and no longer a place where we earn a living (Lowenthal 1997). The link between certain so-called traditional landscape structures and the processes that shaped them has thus been broken (Antrop 2005).

However, re-examining these processes, and, more specifically, the way in which traditional peasant societies "added"[22] to nature, would make it possible to consider other ways of extending agro-pastoral legacies, without necessarily seeking to control landscape forms and by assuming the promotion of new or emerging ecosystems (Hobbs 2006). These approaches necessarily involve the supervision of agricultural practices by rethinking cropping systems in their entirety and not only by greening land use. However, they should also involve a large part of the population so that everyone participates in the management of the environment through their consumption patterns or even by getting directly involved in the maintenance of "natural" spaces. This return to collective actions would at least have the virtue of no longer pitting consumers against producers and of making some people rediscover the pleasures of gardening (in the broadest sense) along with the demands of a life in contact with the elements. It would be yet another way to prolong the traditional agrarian heritage…

Is this realistic? Perhaps no less than "wanting to mummify a space artificially stabilized in time and limited in space" (Bertrand 1975) by drawing up fixed and oversimplified pictures of traditional agrarian landscapes that are supposed to constitute historical baselines for ecological restoration.

22 To use an expression of Maurice Genevois about Pierre de Crescens, the author of *Rustican des labours des champs*, around 1300.

Conclusion

The Baseline Concept: A Narrow Path Between False Trails and True Impasses

We deliberately avoided starting this book with the classic introductory definition by letting the various authors, most of whom are geographers, develop their chapters without a prior framework. In this way, their observations, examples or reflections do not need to correspond to a predetermined vision of the baseline. At the end of this book, the result is a kaleidoscope of texts approaching the baseline notion from different angles and relating to different environments. Such an approach enables us to identify some strong elements concerning the construction and use of this notion for the conservation and management of nature. Thus, we will see that, behind the numerous criticisms formulated against the notion, its use proves to be relevant as long as the false trails and impasses are clearly identified and circumvented. Moreover, speaking of baselines is not only a question of nostalgia, but also of using methods that question the past, as well as the future, space and societies.

False trail: the myth of a single baseline

The first conclusion of this book is to emphasize the diversity of baseline concepts. There is not only one way of conceiving a baseline. It can be a state or a functioning, refer to another site that is geographically very close or several very distant sites, correspond to different periods of time in the past or future, take the form of a scientifically developed biophysical indicator or debate with a plurality of actors, be defined by an exceptional value or an average value of biodiversity or can

---

Chapter written by Simon DUFOUR, Anne-Julia ROLLET, Armelle DECAULNE and Laurent GODET.

correspond to objects varied in nature and scale (an assembly of species, a landscape arrangement, a regime of physical disturbance, etc.). The absence of a single baseline emphasizes that the choice of baseline among several possible ones is a sorting operation, a social construction, and that there is therefore no obvious, natural or normal baseline for a given site. The apparent obviousness of choosing a baseline solely on the basis of its naturalness is undermined by the Anthropocene. In the same way, the baseline situation varies, on the one hand, according to the individuals, their background, scientific education, environment or values, and, on the other hand, according to the interactions between the individuals who make up the group or groups in charge of defining this baseline. The same site can be relatively commonplace for its avifauna, but can present mosses of interest, and can thus represent a baseline for a scientist, but not for their colleague. However, the diversity of criteria for defining a baseline is not a problem in itself. Rather, it demonstrates that the use of baselines is not as caricatured, and that the ambiguity and implicit natures that persist around this notion allow users to shape their own baseline. On the contrary, this raises the question of the modality by which the sorting takes place. Who are the actors who participate? With what power relations? With what cognitive and affective baggage?

A true impasse: when the means becomes the end

The baseline is thus a social construction; there is a risk in not seeing this constructed character and the biases it generates. However, the second conclusion that emerges from this book is that there is probably a more significant risk that could potentially constitute a real impasse: considering a baseline as a management objective when it is in fact a means of action. A baseline is by definition what we refer to, or, in other words, a set of elements that we have chosen or determined beforehand as a framework for situating and solving a problem. It is a selection made within a larger set to establish a relationship between an observed situation and an objective. A functioning, biodiversity, species, structure, set of ecosystem services or landscape are selected in order to compare a given situation and an objective, and establish an evaluation of the state of a given system using this comparison. It is thus a means. Nevertheless, a shift often occurs, and the baseline becomes implicitly desirable, better, etc., *i.e.*, a standard that is imposed as an objective. This is an implicit process because, once this shift has taken place, the constructed character of the baseline, the fact that it is a choice among several possibilities, is forgotten. Transforming a means into an objective is a dead end that can generate misunderstandings, the most emblematic of which is the one that has long crossed (and sometimes still does) the fields of conservation and restoration: the use of knowledge of the past states and functioning of ecosystems as a target for their current management. If these states and functionings are targets, then let us be clear, failure is certain, because reconstituting them in their integrity is impossible. Profound and multiple changes, complex trajectories and successive irreversible processes over long time steps make the past impossible to reconstitute. The case of the forest cover of Iceland, which many consider today as a "land of nature", illustrates the complexity of these trajectories well. Approximately 5–15 million years ago, the forest was common, including redwoods, magnolias and other warm-temperate tree species. At the end of the Pliocene, shortly before the Pleistocene glaciations, the forest type was boreal, with pine, spruce, birch and alder. Successive glaciations severely depleted species diversity; pines survived the early glacial periods, and alder pollen is present in the fossil record of interglacial periods up to 500,000 BP. The only forest species to have reached the present interglacial is downy birch; mountain ash is uncommon, and willows and poplars are extremely rare; all of these species are more likely to be found as shrubs today, with different compositions depending on elevation, shelter positions and distance from the coast. The colonization by humans from 871 promoted the exploitation of wood for heating and construction, the production of charcoal for the manufacture of metal tools, fodder and the clearing of agricultural land. The climatic deterioration of the Little Ice Age and the succession of volcanic eruptions were unfavorable to the spontaneous recovery of afforestation; the failure of the regeneration of birch forests is, however, mainly due to extensive summer grazing of sheep; only the protected forest remnants escaped disappearance, the birch again colonizing the enclosures. Afforestation by plantation began in the second half of the 20th century, and it was at this time that the balance of deforestation and afforestation became favorable again. The law on silviculture and soil conservation enacted in 1907 led to the creation of the forestry service the following year; initially focused on the introduction of exotic species (larch, pine and spruce), efforts were then directed towards the protection of the endemic birch, as well as the Siberian larch and Sitka spruce, whose old stands showed good growth. The financial crisis of 2008–2009 drastically reduced the funding of the Icelandic Forest Service, reducing the volumes of plantations; the collapse of the national currency caused the price of wood imports to rise, which in turn favored the exploitation of thinning in the national forests, now economically profitable. Thus, it is impossible to consider the Icelandic vegetation and forest landscape as "natural" today, and it is especially impossible to reconstruct all the socio-ecological dynamics that have shaped it.

In fact, our relationship with the past is both incomplete (lack of knowledge of the multiple historical trajectories of sites over various time frames) and shifting (the baseline period is a moving window that shifts with the generations and their relationship with their environment). However, if these baselines remain a means, then they enlighten the debate by widening the range of possibilities, activating the imagination and proposing mediums for discussion. For example, understanding the temporal succession of these states enables us to develop a functional vision of the baseline. The trajectory thus reconstructed makes it possible to discuss the dynamics of a site (in transition? in temporary equilibrium? etc.) and avoid giving too much importance to a state defined at a given moment, in a particular context. In turn, this trajectory implies questioning and frequently updating the baselines insofar as they may have been defined on the basis of transient, non-stable conditions (*e.g.*, climate change, changes in land use, etc.).

A narrow path: debating the baselines to aim for better shared choices

The notion of a baseline is sometimes perceived as one of the transformations of a technicist management of nature and its procession of quantitative indicators, cost/benefit analyses, etc. Indeed, if the reference is unique, without any explicit criteria of choice, and becomes an objective, then it necessarily remains open to criticism. For, yes, the arrow of time is unstoppable and it is impossible to find the pre-Anthropocene landscapes. And, yes, space is rough and it is impossible to find a strictly comparable baseline site. However, while baselines remain a means or a tool, a path, a narrow one, opens up. It is an arduous path for managers (and for scientists!), because it is less marked out. On this path, there are many baselines, coming from several approaches: historical reconstitution, spatial comparison, prospective reflection, etc. The baselines serve as reference points and inspiration. For example, it is not possible to revive the Pleistocene megafauna as it was (if we even knew about it), but the biophysical functioning of this period is an element that feeds not only reflection on the range of possibilities, but also the imagination. This can prove to be a very powerful motor for mobilization and action. Similarly, designing a baseline for urban ecosystems may seem counter-intuitive, given that these environments have long been considered as non-natural. However, it becomes possible if we take this approach. Finally, these baselines must be discussed. The existence of discussion and decision-making arenas is a necessary condition for defining management objectives that are as shared as possible. If the objectives are not clearly identified, argued and validated, there is a great risk that the baseline will be substituted for them. The path is arduous because, let us not be naive, this collective definition of objectives is a complex process. It is crossed by struggles between participants (including scientists) who do not have the same roles, powers, objectives, etc. The debate of multiple baselines can be a discussion device, but it does not ensure the quality of the decision process; it is a necessary, however not sufficient condition. Similarly, this debate-based approach does not necessarily exclude the need for recourse to the definition of normative criteria, the mandatory nature of which can also constitute a spur to action.

A geographical look at the notion of a baseline and nature

The chapters gathered in this book also show that the notion of a baseline (*e.g.*, resources, indicators) is a good indicator of the relationship that a society has with nature. Indeed, the diversity of the ways in which a baseline is defined is ultimately a reflection of the ambiguity of the relationship between modern science and the notion of nature. In the image of the nature they are supposed to represent, baselines are both a set of structures and non-human processes that are objectively observable, describable and quantifiable (*e.g.*, a number of species, a volume of dead wood), and at the same time a set that social collectives subjectively decide to distinguish as different and desirable (*e.g.*, a landscape, a service provided). The definition of baselines is thus based on the natural sciences, which characterize states and functions on the basis of biophysical indicators. But it also implies social structures and processes that establish, order and select these indicators and that the social sciences describe and analyze. The example of a baseline thus reminds scientists that they must develop a reflexive look at their own practices in order to understand how their activity fits into a larger social framework.

This duality explains why geography, in addition to other disciplines, helps to shed light on the notion of a baseline. First, it demonstrates that the world is spatially heterogeneous with sites in different locations presenting different biophysical conditions and socio-ecological trajectories. This evidence must be recalled to avoid clumsy transpositions. Each place is unique, and while it is possible, through rigorous descriptions and analyses of numerous sites, to identify similarities and define ranges of values for certain parameters, the fact remains that the intertwining of the links that unite the physical, chemical, biological, human and social dimensions of a place generates a complexity that is difficult to reconcile with simple, universally transferable solutions. Finally, geography is one of the rare disciplines to contain approaches that are rooted in both the natural and social sciences, and which therefore studies both sides of nature.

References

Adams, W.M. and Mulligan, M. (2003). *Decolonizing Nature. Strategies for Conservation in a Post-Colonial Era*. Earthscan, London.

ADEME (2019). Évaluation du gisement relatif aux zones délaissées et artificialisées propices à l'implantation de centrales photovoltaïques. ADEME, Final report, April.

Agnoletti, M. (2014). Rural landscape, nature conservation and culture: Some notes on research trends and management approaches from a (southern) European perspective. *Landscape and Urban Planning*, 126, 66–73.

Aillery, F., Antoni, V., Aouir, C., Arnaud, M., Bardinal, M., Barry, C., Baude, M., Bonnet, A., Besancon, M., Bonnard, P. *et al.* (2018). Environnement & agriculture : les chiffres clés – édition 2018. Commissariat général au développement durable [Online]. Available at: https://www.statistiques.developpement-durable.gouv.fr/sites/default/files/2018-10/datalab-36-environnement-agriculture-les-cc-edition-2018-juin2018.pdf.

Ainsworth, C.H., Pitcher, T.J., Rotinsulu, C. (2008). Evidence of fishery depletions and shifting cognitive baselines in Eastern Indonesia. *Biological Conservation*, 141(3), 848–859.

Alagona, P., Sandlos, J., Wiersma, Y. (2012). Past imperfect: Using historical ecology and baseline data for contemporary conservation and restoration projects. *Environmental Philosophy*, 9(1), 49–70.

Albert, C.H., Hervé, M., Fader, M., Bondeau, A., Leriche, A., Monnet, A.-C., Cramer, W. (2020). What ecologists should know before using land use/cover change projections for biodiversity and ecosystem service assessments. *Regional Environmental Change*, 20, 106.

Albrecht, G. (2006). Solastalgia. *Alternatives Journal*, 32, 34–36.

Alcover, J.A., Sans, A., Palmer, M. (1998). The extent of extinctions of mammals on islands. *Journal of Biogeography*, 25, 913–918.

Allan, J.R., Venter, O., Watson, J.E.M. (2017). Temporally inter-comparable maps of terrestrial wilderness and the Last of the Wild. *Scientific Data*, 4, 170187.

Allen, T.F.H. and Starr, T.B. (1982). *Hierarchy: Perspectives for Ecological Complexity.* University of Chicago Press, Chicago.

Alleway, H.K. and Connell, S.D. (2015). Loss of an ecological baseline through the eradication of oyster reefs from coastal ecosystems and human memory. *Conservation Biology*, 29(3), 795–804.

André, M.F. (1995). Vitesses d'enfrichement de la montagne limousine. *Norois*, 168(1), 629–640.

André, J. (1997). Diversité & gestion écologique : place de la phase hétérotrophe du cycle sylvigénétique. *Ecologie*, 29, 247–25.

Andreu-Boussut, V. (2016). Les salins narbonnais : des espaces en cours de patrimonialisation. In *Ambivalences patrimoniales au Sud. Mise en scène et jeux d'acteurs*, Guillaud, D., Juhé-Beaulaton, D., Cormier-Salem, M.C., Giroud, Y. (eds). IRD Karthala, Paris.

Antrop, M. (2005). Why landscapes of the past are important for the future. *Landscape and Urban Planning*, 70, 21–34.

Aplet, G., Thomson, J., Wilbert, M. (2000). Indicators of wildness: Using attributes of the land to assess the context of wilderness. In *Wilderness Science in a Time of Change Conference – Volume 2: Wilderness within the Context of Larger Systems*, McCool, S.-F., Cole, D.-N., Borrie, W.-T., O'Loughlin, J. (eds). Proceedings RMRS-P-15-VOL-2. U.S. Department of Agriculture, Forest Service, Rocky Mountain Research Station, Ogden, UT.

Apostolopoulou E. and Adams W.M. (2017). Biodiversity offsetting and conservation: Reframing nature to save it. *Oryx*, 51, 23–31.

Arlaud, S. (1994). Friches et jachères en Poitou-Charentes. *Norois*, 164(1), 667–689.

Armanini, D.G., Monk, W.A., Carter, L., Cote, D., Baird, D.J. (2013). Towards generalised reference condition models for environmental assessment: A case study on rivers in Atlantic Canada. *Environmental Monitoring Assessment*, 185, 6247–6259.

Aronson, J., Dhillion, S., Le Floc'h, E. (1995). On the need to select an ecosystem of reference, however imperfect: A reply to Pickett & Parker. *Restoration Ecology*, 3, 1–3.

Aronson, J., Milton, S.J., Blignaut, J.N. (2007). *Restoring Natural Capital: Science, Business, and Practice.* Island Press, Washington, USA.

Baattrup-Pedersen, A., Kristensen, E.A., Jørgensen, J., Skriver, J., Kronvang, B., Andersen, H.E., Hoffman, C.C., Kjellerup Larsen, L.M. (2009). Can a priori defined reference criteria be used to select reference sites in Danish streams? Implications for implementing the Water Framework Directive. *Journal of Environemntal Monitoring*, 11, 344–352.

Bachelard, G. (1957). *La poétique de l'espace.* Presses universitaires de France, Quadrige, Paris.

Baerselman, F. and Vera, F. (1995). *Nature Development. An Exploratory Study for the Construction of Ecological Networks*. Ministry of Agriculture, Nature Management and Fisheries, The Netherlands.

Baiser, B., Olden, J.D., Record, S., Lockwood, J.L., McKinney, M.L. (2012). Pattern and process of biotic homogenization in the New Pangea. *Proceedings of the Royal Society B*, 279, 4772–4777.

Bakkehøi, S., Domaas, U., Lied, K. (1983). Calculation of snow avalanche runout distance. *Annals of Glaciology*, 4, 24–29.

Balaguer, L., Escudero, A., Martín-Duque, J.F., Mola, I., Aronson, J. (2014). The historical reference in restoration ecology: Re-defining a cornerstone concept. *Biological Conservation*, 176, 12–20.

Balmford, A. (2014). *Wild Hope. On the Front Lines of Conservation Success*. The University of Chicago Press, Chicago, London.

Barbault, R. (1997). *Ecologie générale. Structure et fonctionnement de la biosphère*. Masson, Paris.

Barbosa-Filho, M.L., de Souza, G.B.G., de Faria Lopes, S., Siciliano, S., Davis, R.A.H., da Silva Mourão, J. (2020). Evidence of shifting baseline and Fisher judgment on lane snapper (*Lutjanus synagris*) management in a Brazilian marine protected area. *Ocean & Coastal Management*, 183, 105025.

Barkowski, J.W., Kolditz, K., Brumsack, H., Freund, H. (2009). The impact of tidal inundation on salt marsh vegetation after de-embankment on Langeoog Island, Germany. Six years time series on permanent plots. *Journal of Coastal Conservation*, 13, 185–206.

Barnaud, C. and Couix, N. (2020). The multifunctionality of mountain farming: Social constructions and local negotiations behind an apparent consensus. *Journal of Rural Studies*, 73, 34–45.

Barnaud, C. and Mathevet, R. (2015). Géographie et participation : des relations complexes et ambigues. In *Pour une géographie de la conservation. Biodiversités, natures et sociétés*, Mathevet, R. and Godet, L. (eds). L'Harmattan, Paris.

Barnaud, C., Fischer, A., Staddon, S., Blackstock, K., Moreau, C., Corbera, E., Hester, A., Mathevet, R., McKee, A., Reyes, J. *et al.* (2021). Is forest regeneration good for biodiversity? Exploring the social dimensions of an apparently ecological debate. *Environmental Science and Policy*, 120, 63–72.

Barnosky, A.D., Matzke, N., Tomiya, S., Wogan, G.O., Swartz, B., Quental, T.B., Marshall, C., McGuire, J.L., Lindsey, E.L., Maguire, K.C. *et al.* (2011). Has the Earth's sixth mass extinction already arrived? *Nature*, 471, 51–57.

Baron-Yellès, N. (2000). *Recréer la nature. Écologie, paysage et société au marais d'Orx*. ENS éditions, Paris.

Barraud, R. (2020). Les références spatiales et temporelles des paysages forestiers du rewilding en Europe : imaginaires, discours et projets. *Projets de paysage* [Online]. Available at: http://journals.openedition.org/paysage/9076 [Accessed 24 July 2020].

Barraud, R. (2021). Les références spatiales et temporelles des paysages forestiers du *rewilding* en Europe : imaginaires, discours et projets. *Projets de paysage*, 22 [Online]. Available at: http://journals.openedition.org/paysage/9076 [Accessed 26 February 2021].

Barraud, R., Andreu-Boussut, V., Chadenas, C., Portal, C., Guyot, S. (2019). Ensauvagement et ré-ensauvagement de l'Europe : controverse et postures scientifiques. *Bulletin de l'Association de géographes français*, 96(2), 301–318.

Barthod, C. and Trouvilliez, J. (2002). La protection des forêts dans la politique forestière française, le cas particulier des réserves intégrales. *Revue forestière française*, 54(1), 7–16.

Bartlett, L.J., Williams, D.R., Prescott, G.W., Balmford, A., Green, R.E., Eriksson, A., Valdes, P.J., Singarayer, J.S., Manica, A. (2015). Robustness despite uncertainty: Regional climate data reveal the dominant role of humans in explaining global extinctions of Late Quaternary megafauna. *Ecography*, 38, 1–10.

Bas, C., Imbert, I., Clermont, S., Reinert, M.E., Calvet, C., Berté, C., Vaissière, A.C. (2019). Approches anticipées et planifiées de la compensation écologique en Allemagne : vers un retour d'expérience pour la France ? *Sciences, Eaux & Territoires*, 31, 44–48.

Basset, K.L. (2009). *Aux origines du Parc national des Cévennes. Histoire des parcs nationaux*. Éditions Quae, Versailles.

Bastien, Y. and Gauberville, C. (2011). *Vocabulaire forestier, écologie, gestion et conservation des espaces boisés*. Institut pour le développement forestier, AgroParisTech, Office national des forêts, Paris.

Baum, J.K. and Myers, R.A. (2004). Shifting baselines and the decline of pelagic sharks in the Gulf of Mexico. *Ecology Letters*, 7(2), 135–145.

Bella, E.A., Patrick Boehnkea, T., Harrisona, M., Maob, W.L. (2015). Potentially biogenic carbon preserved in a 4.1 billion-year-old zircon. *Proceedings of the National Academy of Sciences of the USA*, 112, 14518–14521.

Belyazid, S., Sverdrup, H., Kurz, D., Braun, S. (2011). Exploring ground vegetation change for different deposition scenarios and methods for estimating critical loads for biodiversity using the forsafe-veg model in Switzerland and Sweden. *Water Air and Soil Pollution*, 216(1–4), 289–317.

Bennett, C.F. (1975). *Man and Earth's Ecosystems*. John Wiley and Sons, New York.

Bépoix, S. and Richard, H. (2019). *La forêt au Moyen Âge*. Les Belles Lettres, Paris.

Berger, G. (1958). L'attitude prospective. *Revue Prospective*, 1.

Bernard, M. (2018). Changements climatiques et herbivorie : influence sur la régénération et le potentiel d'avenir des forêts mélangées. PhD Thesis, Université de Montpellier, Montpellier.

Bernard, V., Billard, C., Couturier, Y., Jaouen, G., LeDigol, Y., Mélin, M., Mougne, C. (2012). Quand nos ancêtres allaient au pieu : des chaines de production forestière du bronze ancien tournées vers le taillis. *Actes du colloque L'Homme, ses ressources et son environnement, dans le Nord-Ouest de la France à l'âge du Bronze : actualités de la recherche*, Éditions de Géosciences, Rennes, 27–58.

Bernardie, S., Vandromme, R., Thiery, Y., Houet, T., Grémont, M., Masson, F., Grandjean, G., Bouroullec, I. (2021) Modelling landslide hazards under global changes: The case of a Pyrenean valley. *Natural Hazards and Earth System Sciences*, 21(1).

Bertrand, G. (1975). Pour une histoire écologique de la France rurale. In *Histoire de la France rurale*, Duby, G. and Wallon, A. (eds). Le Seuil, Paris.

Bezombes, L. (2017). Développement d'un cadre méthodologique pour l'évaluation de l'équivalence écologique : application dans le contexte de la séquence "Éviter, Réduire, Compenser" en France. Ecology PhD Thesis, Université de Grenoble Rhône-Alpes.

Bezombes, L., Gaucherand, S., Kerbiriou, C., Reinert, M.-E., Spiegelberger, T. (2017). Ecological equivalence assessment methods: What trade-offs between operationality, scientific basis and comprehensiveness? *Environmental Management,* 60, 216–230.

Bezombes, L., Gaucherand, S., Spiegelberger, T., Gouraud, V., Kerbiriou, C. (2018). A set of organized indicators to conciliate scientific knowledge, offset policies requirements and operational constraints in the context of biodiversity offsets. *Ecological Indicators*, 93, 1244–1252.

Bezombes, L., Kerbiriou, C., Spiegelberger, T. (2019). Do biodiversity offsets achieve no net loss? An evaluation of offsets in a French department. *Biological Conservation*, 231, 24–29.

Bigard, C., Pioch, S., Thompson, J.D. (2017). The inclusion of biodiversity in environmental impact assessment: Policy-related progress limited by gaps and semantic confusion. *Journal of Environmental Management*, 200, 35–45.

Bioret, F. and Chlous-Ducharme, F. (2011). Évaluer la dégradation en écologie de la restauration, une question d'échelles de références et de perception. *Sciences eaux & territoires*, 5, 3–5.

Bioret, F., Gourmelon, F., Le Berre, I. (1994). Analyse spatiale du processus d'enfrichement sur l'île d'Ouessant (Finistère). *Norois*, 164(1), 547–558.

Biotope and DREAL LR (2016). Vers une harmonisation des pratiques d'évitement, de réduction et de compensation des impacts sur la biodiversité, application à la Région Occitanie. Report, Biotope and DREAL LR.

Birks, H.J.B. and Birks, H.H. (1980). *Quaternary Palaeoecology*. Edward Arnold, London.

Blandin, P. (2009). *De la protection de la nature au pilotage de la biodiversité*. Éditions Quae, Versailles.

Blandin, P. (2014). Est-il juste de manipuler la nature ? In *Ingénierie écologique. Action par et/ou pour le vivant*, Rey, F. (ed.). Éditions Quae, Versailles.

Blanqui, A. (1851). Tableau des populations rurales de la France. *Journal des économistes*, 30, 1–15.

Blicharska, M., Angelstam, P., Giessen, L., Hilszczański, J., Hermanowicz, E., Holeksa, J., Jacobsen, J.B., Jaroszewicz, B., Konczal, A., Konieczny, A. *et al.* (2020). Between biodiversity conservation and sustainable forest management – A multidisciplinary assessment of the emblematic Białowieża Forest case. *Biological Conservation*, 248, 108614.

Block, M. (1860). *Statistique de la France – comparée avec les autres Etats de l'Europe*. Lib. d'Aymiot, T.2, Paris.

Blondel, J. (2020). *Biodiversité. Un nouveau récit à écrire*. Éditions Quae, Paris.

Bobbink, R., Hicks, K., Galloway, J., Spranger, T., Alkemade, R., Ashmore, M., Bustamante, M., Cinderby, S., Davidson, E., Dentener, F. *et al.* (2010). Global assessment of nitrogen deposition effects on terrestrial plant diversity: A synthesis. *Ecological Applications*, 20, 30–59.

Bobiec, A., Gutowski, J., Zub, K., Pawlaczyk, P., Laudenslayer, W. (2005). *The Afterlife of a Tree*. WWF Poland, Warszawa-Hajnowka.

Boileau, J., Calvet, C., Pioch, S., Moulhérat, S. (2022). Ecological equivalence assessment: The potential of genetic tools, remote sensing and ecological modelling to better apply the mitigation hierarchy. *Journal of Environmental Management*, 305, 114415.

Boisvert, V. (2015). Conservation banking and the economization of nature: An institutional analysis. *Ecosystem Services*, 15, 134–142.

Boisvert, V., Méral, P., Froger, G. (2013). Market-based instruments for ecosystem services: Institutional innovation or renovation? *Society and Natural Resources*, 26, 1122–1136.

Bonneau, M. and Barbaza, Y. (1974). Conflits d'utilisation du sol et besoins en espace récréatif : le cas des grandes prairies angevines. *Bulletin de l'Association de Géographes Français*, 417–418, 185–193.

Bonneuil, C. (2015). Tell me where you come from, I will tell you who you are: A genealogy of biodiversity offsetting mechanisms in historical context. *Biological Conservation*, 192, 485–491.

Bonneuil, C. (2017). Capitalocène. *EcoRev*, 44, 52–60.

Bonneuil, C. and Fressoz, J.B. (2013). *L'événement anthropocène*. Le Seuil, Paris.

Bontemps, J.D., Herve, J.C., Duplat, P., Dhote, J.-F. (2012). Shifts in the height-related competitiveness of tree species following recent climate warming and implications for tree community composition: The case of common beech and sessile oak as predominant broadleaved species in Europe. *Oikos*, 121(8), 1287–1299.

Bouleau, G. and Pont. D. (2015). Did you say reference conditions? Ecological and socio-economic perspectives on the European water framework directive. *Environmental Science and Policy*, 47, 32–41.

Bowman, M.F. and Somers, K.M. (2006). Evaluating a novel test site analysis (TSA) bioassessment approach. *Journal of the North American Benthological Society*, 25, 712–727.

Bradshaw, R.H., Hannon, G.E., Lister, A.M. (2003). A long-term perspective on ungulate–vegetation interactions. *Forest Ecology and Management*, 181, 267–280.

Brahic, E. and Rambonilaza, T. (2016). Quelle valeur les français accordent-ils à la biodiversité des forêts métropolitaines ? *Etudes et documents*, 141, Commissariat général au développement durable, Paris.

Brattstrom, B.H. (1990). Biogeography of the Islas Revillagigedo, Mexico. *Journal of Biogeography*, 17, 177–183.

Brierley, G.J. and Fryirs, K.A. (2005). *Geomorphology and River Management: Applications of the River Styles Framework.* Blackwell Publishing, Malden.

Brousse, G., Arnaud-Fassetta, G., Liébault, F., Bertrand, M., Melun, G., Loire, R., Malavoi, J.R., Fantino, G., Borgniet, L. (2019). Channel response to sediment replenishment in a large gravel-bed river: The case of the Saint-Sauveur dam in the Buëch River (Southern Alps, France). *River Research and Applications*, 36(6), 880–893.

Brown, A.G., Lespez, L., Sear, D.A., Macaire, J.-J., Houben, P., Klimek, K., Brazier, R.E., Van Oost, K., Pears, B. (2018). Natural vs anthropogenic streams in Europe: History, ecology and implications for restoration, river-rewilding and riverine ecosystem services. *Earth-Science Reviews*, 180, 185–205.

Brownlie, S., King, N., Treweek, J. (2013). Biodiversity tradeoffs and offsets in impact assessment and decision making: Can we stop the loss? *Impact Assessment and Project Appraisal*, 31, 24–33.

Bruciamacchie, M. (2009). Biodiversité, naturalité, résilience et plasticité : nouveaux concepts au service de la gestion des arbres. *Revue forestière française*, LXI, 5, 441–446.

Brugnot, G. (2002). Développement des politiques forestières et naissance de la restauration des terrains de montagne. *Annales des ponts et chaussées*, 23–30.

Brunner, S.H., Huber, R., Grêt-Regamey, A. (2016). A backcasting approach for matching regional ecosystem services supply and demand. *Environmental Modelling & Software*, 75, 439–458 [Online]. Available at: https://doi.org/10.1016/j.envsoft.2015.10.018.

Bull, J.W. and Strange, N. (2018). The global extent of biodiversity offset implementation under no net loss policies. *Nature Sustainability*, 1, 790–798.

Bull, J.W., Gordon, A., Law, E.A., Suttle, K.B., Milner-Gulland, E.J. (2014). Importance of baseline specification in evaluating conservation interventions and achieving no net loss of biodiversity. *Conservation Biology*, 28, 799–809.

Bunce, M., Rodwell, L.D., Gibb, R., Mee, L. (2008). Shifting baselines in fishers' perceptions of island reef fishery degradation. *Ocean & Coastal Management*, 51(4), 285–302.

Burbano, D.V., Mena, C.F., Guarderas, P., Vinueza, L., Reck, G. (2014). Shifting baselines in the Galapagos white fin fishery, using fisher's anecdotes to reassess fisheries management: The case of the Galapagos grouper. *The Galapagos Marine Reserve*, Denkinger, J. and Vinueza, L. (eds). Springer, Cham.

Burrascano, S., Keeton, W.S., Sabatini, F.M., Blasi, C. (2013). Commonality and variability in the structural attributes of moist temperate old-growth forests: A global review. *Forest Ecology and Management*, 291, 458–479.

Business and Biodiversity Offstes Progamme (BBOP) (2012). Standard on Biodiversity Offsets. BBOP, Washington, DC.

Büttner, G., Maucha, G., European Environment Agency (2006). The thematic accuracy of Corine land cover 2000 assessment using LUCAS (land use cover area frame statistical survey). EEA Technical report, No. 7/2006.

Callicott, J.B. (2002). Choosing appropriate temporal and spatial scales for ecological restoration. *Journal of Bioscience*, 27, 409–420.

Callicott, J.B., Crowder, L.B., Mumford, K. (1999). Current normative concepts in conservation. *Conservation Biology*, 13(1), 22–35.

Calvet, C., Ollivier, G., Napoléone, C. (2015a). Tracking the origins and development of biodiversity offsetting in academic research and its implications for conservation: A review. *Biological Conservation*, 192, 492–503.

Calvet, C., Levrel, H., Napoléone, C., Dutoit, T. (2015b). La réserve d'actifs naturels : une nouvelle forme d'organisation pour la préservation de la biodiversité en France ? In *Restaurer la nature pour atténuer les impacts du développement. Analyse des mesures compensatoires pour la biodiversité*, Levrel, H., Frascaria, N., Hay, J., Martin, G., Pioch, S. (eds). Éditions Quae, Versailles.

Calvet, C., Napoléone, C., Salles, J.-M. (2015c). The biodiversity offsetting dilemma: Between economic rationales and ecological dynamics. *Sustainability*, 7, 7357–7378.

Calvet, C., Delbar, V., Chapron, P., Brasebin, M., Perret, J., Moulherat, S. (2019a). La biodiversité à l'épreuve des choix d'aménagement : une approche par la modélisation appliquée à la région Occitanie. *Sciences, eaux & territoires*, 31, 24–31.

Calvet, C., Le Coënt, P., Quétier, F., Napoleone, C. (2019b). Challenges of achieving biodiversity offsetting through agri-environmental schemes: Evidence from an empirical study. *Ecological Economics*, 163, 113–125.

Campbell, L.M., Gray, N.J., Hazen, E.L., Shackeroff, J.M. (2009). Beyond baselines: Rethinking priorities for ocean conservation. *Ecology and Society*, 14(1).

Canfield, D.E., Glazer, A.N., Falkowksi, P.G. (2010). The evolution and future of Earth's nitrogen cycle. *Science*, 330, 192–196.

Cantelaube, P. and Carles, M. (2014). Le registre parcellaire graphique : des données géographiques pour décrire la couverture du sol agricole. *Le Cahier des Techniques de l'INRA (Numéro Spécial Géo Expé)*, 58–64.

Carnino, N. and Touroult, J. (2010). Évaluation de l'état de conservation des habitats forestiers à l'échelle d'un site Natura 2000 : du concept vers un outil pour le gestionnaire. *Revue forestière française*, 62(2), 127–140.

Caro, T., Darwin, J., Forrester, T., Ledoux-Bloom, C., Wells, C. (2011). Conservation in the Anthropocene. *Conservation Biology*, 26(1), 185–188.

Carver, S. (1991). Integrating multi-criteria evaluation with geographical information systems. *International Journal of Geographical Information Systems*, 5, 321–339.

Carver, S. (1996). Mapping the wilderness continuum using raster GIS. In *Raster Imagery in Geographic Information Systems*, Morain, S. and Lopez-Baros, S. (eds). OnWord Press, New Mexico.

Carver, S. (2019). Rewilding through land abandonment. In *Rewilding*, 1st edition, Pettorelli, N., Durant, S.M., du Toit, J.T. (eds). Cambridge University Press, Cambridge.

Carver, S. and Fritz, S. (1999). Mapping remote areas using GIS. In *Landscape Character: Perspectives on Management and Change*, Usher, M. (ed.). Natural Heritage of Scotland Series, HMSO.

Carver, S., Evans, A., Fritz, S. (2002). Wilderness attribute mapping in the United Kingdom. *International Journal of Wilderness*, 8, 24–29.

Carver, S., Comber, A., McMorran, R., Nutter, S. (2012). A GIS model for mapping spatial patterns and distribution of wild land in Scotland. *Landscape and Urban Planning*, 104, 395–409.

Carver, S., Tricker, J., Landres, P. (2013). Keeping it wild: Mapping wilderness character in the United States. *Journal of Environmental Management*, 131, 239–255.

Castillo, C.P. (2021). Modelling agricultural land abandonment in a fine spatial resolution multi-level land-use model: An application for the EU. *Environmental Modelling and Software*, 2021, 27.

Castree, N. (2003). Commodifying what nature? *Progress in Human Geography*, 27(3), 273–297.

Cateau, E., Larrieu, L., Vallauri, D., Savoie, J.M., Touroult, J., Brustel, H. (2015). Ancienneté et maturité : deux qualités complémentaires d'un écosystème forestier. *Comptes rendus biologies*, 338(1), 58–73.

Cazalis, F. and Granon, S. (2017). Faire face aux nouveaux défis environnementaux. Comment notre cerveau permet les changements de nos comportements. In *Le souci de la nature*, Fleury, C. and Prévot A.C. (eds). CNRS Editions, Paris.

CBD and UNEP (2010). Biodiversity offsets: A tool for CBD parties to consider and a briefing on the business and biodiversity offsets programme (BBOP). *Conference of the Parties to the Convention on Biological Diversity*, Tenth Meeting, Nagoya, Japan.

Ceaușu, S., Hofmann, M., Navarro, L.M., Carver, S., Verburg, P.H., Pereira, H.M. (2015). Mapping opportunities and challenges for rewilding in Europe. *Conservation Biology*, 29, 1017–1027.

Ceballos, G., Garcia, A., Ehrlich, P.R. (2010). The sixth extinction crisis: Loss of animal populations and species. *Journal of Cosmology*, 8, 1821–1831.

Ceballos, G., Ehrlich, P.R., Barnosky, A.D., García, A., Pringle, R.M., Palmer, T.M. (2015). Accelerated modern human–induced species losses: Entering the sixth mass extinction. *Science Advances*, 1, 1–5.

CEREMA (2017). Rapport d'analyse – enquête nationale sur les Observatoires Régionaux d'espaces naturels, agricoles et forestiers. Report, CEREMA.

Chabrerie, O., Loinard, J., Perrin, S., Saguez, R., Decocq, G. (2010). Impact of prunus serotina invasion on understory functional diversity in a European temperate forest. *Biological Invasions*, 12(6), 1891–1907.

Chadenas, C. (2008). *L'Homme et l'oiseau sur les littoraux d'Europe occidentale.* L'Harmattan, Paris.

Chadenas, C., Andreu-Boussut, V., Barthon, C., Michel, X., de Lajartre, A. (2020). La gouvernance des espaces protégés : vers un partage de la nature ? *VertigO – la revue électronique en sciences de l'environnement*, 20(1) [Online]. Available at: http://journals.openedition.org/vertigo/27491 [Accessed 22 February 2021].

Chapron, G., Kaczensky, P., Linnel, J.D.C., von Arx, M., Huber, D., Andrén, H., López-Bao, J.V., Adamec, M., Álvares, F., Anders, O. *et al.* (2014). Recovery of large carnivores in Europe's modern human-dominated landscapes. *Science*, 346, 1517–1519.

Charaud, A.-M. (1949). Bocage et plaine dans l'Ouest de la France. *Annales de Géographie*, 310, 113–125.

Chassé, P., Blatrix, C., Frascaria-Lacoste, N.F. (2020). What is wrong between ecological science and policy? *Ecology Letters*, 23, 1736–1738.

Chazal, P. and Dumont, R. (1954). La nécessaire révolution fourragère et l'expérience lyonnaise. *Journal de la France agricole*, Paris.

Chevalier, A. (1949). Les haies plantées d'arbres des pays de bocage ont-elles leur raison d'être ? *Comptes rendus des séances de l'Académie d'agriculture de France*, 5, 149–152.

Chevassus-au-Louis, B., Salles, J.M., Pujol, J.L. (2009). *Approche économique de la biodiversité et des services liés aux écosystèmes.* Contribution à la décision publique, Conseil d'analyse stratégique, Paris.

Chibilev, A.A. (2017). The origins and development paths of zapovednik management in Russia. *Geography and Natural Resources*, 3, 6–12.

Chiffaut, A. (2006). *Guide méthodologique des plans de gestion de réserves naturelles*, 79. MEED, ATEN, Montpellier.

Chouquer, G. (2007). Le paysage ou la mémoire des formes. *Cosmopolitiques n°15. Esthétique et espace*, 43–52.

Christensen, M., Hahn, K., Mountford, E.P., Odor, P., Standovar, T., Rozenbergar, D., Diaci, J., Wijdeven, S., Meyer, P., Winter, S. *et al.* (2005). Dead wood in European beech (*Fagus sylvatica*) forest reserves. *Forest Ecology and Management*, 210(1–3), 267–282.

Ciais, P. and Sabine, C. (eds) (2013). Carbon and other biogeochemical cycles. In *Climate Change 2013: The Physical Science Basis Contribution of Working Group I to the Fifth Assessment Report of the Intergovernmental Panel on Climate Change*, Stocker, T.F., Qin, D., Plattner, G.-K., Tignor, M., Allen, S.K., Boschung, J., Nauels, A., Xia, Y., Bex, V., Midgley, P.M. (eds). Cambridge University Press, Cambridge.

Clair, M., Gaudillat, V., Michez, N., Poncet, R. (2019). HABREF v5.0, référentiel des typologies d'habitats et de végétation pour la France. Guide méthodologique. Report, UMS PatriNat (AFB-CNRS-MNHN), Paris.

Clément, G. (2004). *Manifeste du Tiers paysage.* Sujet/Objet, Paris.

Clement, S. (2021). *Governing the Anthropocene: Novel Ecosystems, Transformation and Environmental Policy.* Palgrave Macmillan, Cham.

Clements, F.E. (1916). *Plant Succession: An Analysis of the Development of Vegetation.* Carnegie Institution Publication, No. 42, Washington, DC.

Clewell, A. and Aronson, J. (2010). *La restauration écologique. Principes, valeurs et structure d'une profession émergente.* Actes Sud, Arles.

Closset-Kopp, D. (2004). Sylvigenèse de la hêtraie-sapinière dans le contexte vosgien. PhD Thesis, Université de Metz, Metz.

Cochet, G. and Durand, S. (2018). *Ré-ensauvageons la France : plaidoyer pour une nature sauvage et libre.* Actes sud, Arles.

Cochet, G. and Kremer-Cochet, B. (2020). *L'europe réensauvagée : vers un nouveau monde.* Actes sud, Arles.

Cohen, K.M., Finney, S.C., Gibbard, P.L., Fan, J.-X. (2013). The ICS International Chronostratigraphic Chart. *Episodes*, 36, 199–204.

Collomb, G. (1977). Les agriculteurs, jardiniers de la nature ? Le cas du Beaufortin. *Études rurales*, 66, 37–42.

Comber, A., Carver, S., Fritz, S., McMorran, R., Washtell, J., Fisher, P. (2010). Different methods, different wilds: Evaluating alternative mappings of wildness using fuzzy MCE and Dempster-Shafer MCE. *Computers, Environment and Urban Systems*, 34, 142–152.

Combes, F. (1989). Restauration des terrains en montagne. Du rêve à la réalité. *Revue forestière française*, 41, 91–106.

Commission Européenne (2009). *Guide sur la chasse durable en application de la directive oiseaux.* Directive 79/409/CEE du Conseil concernant la conservation des oiseaux sauvages.

Conedera, M., Colombaroli, D., Tinner, W., Krebs, P., Whitlock, C. (2017). Insights about past forest dynamics as a tool for present and future forest management in Switzerland. *Forest Ecology and Management*, 388, 100–112.

Corbelle-Rico, E., Crecente-Maseda, R., Santé-Riveira, I. (2012). Multi-scale assessment and spatial modelling of agricultural land abandonment in a European peripheral region: Galicia (Spain), 1956–2004. *Land Use Policy*, 29(3), 493–501.

Cordonnier, T. (2004). Perturbations, diversité et permanences des structures dans les écosystèmes forestiers. PhD Thesis, Université Paris XI, Paris.

Corlett, R. (2013). The shifted baseline: Prehistoric defaunation in the tropics and its consequences for biodiversity conservation. *Biological Conservation*, 163, 13–21.

Croll, D.A., Maron, J.L., Estes, J.A., Danner, E.M., Byrd, G.V. (2005). Introduced predators transform subarctic islands from grassland to tundra. *Science*, 307, 1959–1961.

Cronon, W. (1996). The trouble with wildnerness: Or, getting back to the wrong nature. *Environmental History*, 1, 7–28.

Crooks, K.R. and Soulé, M.E. (1999). Mesopredator release and avifaunal extinctions in a fragmented system. *Nature*, 400, 563–566.

Crosby, A.W. (2003). *The Columbian Exchange: Biological and Cultural Consequences of 1492*. Praeger, Westport.

Cross, J. (1981). The establishment of rhododendron ponticum in the Killarney oakwoods, SW Ireland. *The Journal of Ecology*, 69(3), 807–824.

Crowley, B.E. (2010). A refined chronology of prehistoric Madagascar and the demise of the megafauna. *Quaternary Science Reviews*, 29, 2591–2603.

Crutzen, P.J. (2002). Geology of mankind. *Nature*, 415, 23.

Crutzen, P.J. and Stoermer, E.F. (2000). The Anthropocene. *IGBP Global Change News*, 41, 17–18.

Da Lage, A. and Métaillé, G. (2000). *Dictionnaire de biogéographie végétale*. CNRS éditions, Paris.

Dagognet, F. (1982). *Mort du paysage ? Philosophie et esthétique du paysage*. Ed. Champ Vallon, Milieux.

Dallas, H.F. (2013). Ecological status assessment in Mediterranean rivers: Complexities and challenges in developing tools for assessing ecological status and defining reference conditions. *Hydrobiologia*, 719, 483–507.

Dantec, R. (2017). Rapport sur la réalité des mesures de compensation des atteintes à la biodiversité engagées sur des grands projets d'infrastructures, intégrant les mesures d'anticipation, les études préalables, les conditions de réalisation et leur suivi, Tome 1. Report, Sénat, Paris.

Darwin, C. (1859). *On the Origin of Species by Mean of Natural Selection or the Preservation of Favoured Races in the Struggle for Life*. D. Appleton and Company, New York.

Dauguet, B. (2015). Biodiversity offsetting as a commodification process: A French case study as a concrete example. *Biological Conservation*, 192, 533–540.

Davasse, B. (2015). La trace des temps. Les complexes socio-écologiques au prisme du paysage. Pour une géographie de l'environnement impliquée. HDR Thesis, Université de Toulouse Jean Jaurès.

Davodeau, H. and Barraud, R. (2018). Quelle position adopter face au patrimoine qui vient ? La médiation paysagère pour mettre en débat les temporalités. *Développement durable et territoires. Économie, géographie, politique, droit, sociologie*, 9(2).

Dawson, C. and Hendee, J.C. (2009). *Wilderness Management: Stewardship and Protection of Resources and Values*, 4th edition. Fulcrum, Golden.

Decaulne, A. (2016). Lichenometry in Iceland, results and application. *Géomorphologie : relief, processus, environnement*, 22(1), 77–91.

Delamarre, A. (2002) *La prospective territoriale*. La Documentation française, Paris.

Delparte, D. (2008). Avalanche terrain modeling in Glacier National Park, Canada. Doctoral Thesis, Department of Geography, University of Calgary.

Demangeot, J. (1984). *Les milieux "naturels" du globe*. Masson, Paris.

Deryabina, T.G., Kuchmel, S.V., Nagorskaya, L.L., Hinton, T.G., Beasley, J.C., Lerebours, A., Smith, J.T. (2015). Long-term census data reveal abundant wildlife populations at Chernobyl. *Current Biology*, 25(19), R824–R826.

Descola, P. (2015). Humain, trop humain. *Esprit*, 12, 8–22.

Devictor, V. (2018). La compensation écologique : fondements épistémiques et reconfigurations technoscientifiques. *Natures Sciences Sociétés*, 26(2), 136–149.

Devoucoux, P., Besnard, A., Bretagnolle, V. (2019). Sex-dependent habitat selection in a high-density Little Bustard Tetrax tetrax population in southern France, and the implications for conservation. *Ibis*, 161, 310–324.

Dietre, B., Reitmaier, T., Walser, C., Warnk, T., Unkel, I., Hajdas, I., Lambers, K., Reidl, D., Haas, J.N. (2020). Steady transformation of primeval forest into subalpine pasture during the late Neolithic to Early Bronze Age (2300–1700 BC) in the Silvretta Alps, Switzerland. *The Holocene*, 30(3), 355–368.

Dirzo, R., Young, H.S., Galetti, M., Ceballos, G., Isaac, N.J.B., Collen, B. (2014). Defaunation in the Anthropocene. *Science*, 345, 401–406.

Doak, D.F., Bakker, V.J., Goldstein, B.E., Hale, B. (2013). What is the future of conservation? *Trends in Ecology and Evolution*, 29, 77–81.

Dodd, M.S., Papineau, D., Grenne, T., Slack, J.F., Rittner, M., Pirajno, F., O'Neil, J., Crispin, T.S. (2017). Little evidence for early life in Earth's oldest hydrothermal vent precipitates. *Nature*, 543, 60–64.

Donadieu, P. (2002). Les références en écologie de la restauration. *Revue d'écologie*, 9(supplément), 109–120.

Donald, P.F., Green, R.E., Heath, F.M. (2001). Agricultural intensification and the collapse of Europe's farmland bird populations. *Proceedings of the Royal Society – B*, 268, 25–29.

Donlan, J., Berger, J., Bock, C.E., Bock, J.H., Burney, D.A., Estes, J.A., Foreman, D., Martin, P.S., Roemer, G.W., Smith, F.A. *et al.* (2005). Re-wilding North America. *Nature*, 436, 913–914.

Doughty, C.E., Wolf, A., Field, C.B. (2010). Biophysical feedbacks between the Pleistocene megafauna extinction and climate: The first human-induced global warming? *Geophysical Research Letters*, 37, L15703.

Doussan, I. (2018). Quand les parlementaires débattent de la compensation écologique : des occasions manquées. *Natures Sciences Sociétés*, 26(2), 159–169.

Downs, P.W., Singer, M.S., Orr, B.K., Diggory, Z.E., Church, T.C. (2011). Restoring ecological integrity in highly regulated rivers: The role of baseline data and analytical references. *Environmental Management*, 48, 847–864.

Drapier, N. (2005). Les réserves forestières intégrales, objet d'étude pour les sciences humaines et sociales ? In *La nature a-t-elle encore sa place dans les milieux géographiques*, Arnould, P. and Glon, E. (eds). Publication de la Sorbonne, Paris.

Drew, J., Philipp, C., Westneat, M.W. (2013). Shark tooth weapons from the 19th century reflect shifting baselines in Central Pacific predator assemblies. *PLoS One*, 8(4), e59855.

Dubois, P.J. (2012). *La grande Amnésie écologique*. Delachaux et Niestlé, Paris.

Dubois, J.-J., Hotyat, M., Courtois, N. (2005). Les réserves biologiques en France : de la dynamique dirigée à la dynamique spontanée. In *La nature a-t-elle encore sa place dans les milieux géographiques*, Arnould, P. and Glon, E. (eds). Publication de la Sorbonne, Paris.

Dudgeon, D. (2010). Prospects for sustaining freshwater biodiversity in the 21st century: Linking ecosystem structure and function. *Current Opinion in Environmental Sustainability*, 2(5–6), 422–430.

Dull, R.A., Nevle, R.J., Wood, W.I., Bird, D.K., Avnery, S., Denevan, W.M. (2010). The Columbian encounter and the little ice age: Abrupt land use change, fire, and greenhouse forcing. *Annals of the Association of American Geographers*, 100, 755–771.

Dupouey, J.L., Dambrine, E., Laffite, J.D., Moares, C. (2002). Irreversible impact of past land use of forest soils and biodiversity. *Ecology*, 83, 2978–2984.

Eckersley, R. (2017). La démocratie à l'ère de l'Anthropocène. *La pensée écologique*, 2017-1.

Edwards, G.P., Eldridge, S.R., Wurst, D., Berman, D.M., Garbin, V. (2001). Movement patterns of female feral camels in central and northern Australia. *Wildlife Research*, 28, 283–289.

Elias, S. and Schreve, D. (2007). Late Pleistocene megafaunal extinctions. In *Encyclopedia of Quaternary Science*, Volume 4, Elias, S.A. (ed.). Elsevier, Amsterdam.

Elissalde, B. (2000). Géographie, temps et changement spatial. *Espace géographique*, 29, 224–236.

Ellis, E.C. and Ramankutty, N. (2008). Putting people in the map: Anthropogenic biomes of the world. *Frontiers in Ecology and the Environment*, 6, 439–447.

Ellis, E.C., Gauthier, N., Klein Goldewijk, K., Bliege Bird, R., Boivin, N., Díaz, S., Fuller, D.Q., Gill, J.L., Kaplan, J.O., Kingston, N. *et al.* (2021). People have shaped most of terrestrial nature for at least 12,000 years. *Proceedings of the National Academy of Sciences of the USA*, 118(17), e2023483118.

Emsellem, K., Liziard, S., Scarella, F. (2012). La géoprospective : l'émergence d'un nouveau champ de recherche ? *L'Espace géographique*, 2(2), 154–168.

Enoul, P. (1999). Restaurer nos espaces naturels. Bilan d'un savoir-faire en Bretagne. Institut régional du patrimoine de Bretagne, Rennes.

Estel, S., Kuemmerle, T., Alcántara, C., Levers, C., Prishchepov, A., Hostert, P. (2015). Mapping farmland abandonment and recultivation across Europe using MODIS NDVI time series. *Remote Sensing of Environment*, 163, 312–325.

Estes, J.A. and Duggins, D.O. (1995). Sear others and kelp forests in Alaska: Generality and variation in a community ecological paradigm. *Ecological Monographs*, 65, 75–100.

Estes, J.A., Terborgh, J., Brashares, J.S., Power, M.E., Berger, J., Bond, W.J., Carpenter, S.R., Essington, T.E., Holt, R.D., Jackson, J.B.C. *et al.* (2011). Trophic downgrading of planet Earth. *Science*, 333, 301–306.

Esteves, L. (2014). *Managed Realignment: A Viable Long-Term Coastal Management Strategy?* Springer, Dordrecht.

Étienne, M. (2010). *La modélisation d'accompagnement : une démarche participative en appui au développement durable*. Éditions Quae, Versailles.

Etienne, P. and Lauzet J. (2009). *L'Ours brun. Biologie et histoire, des Pyrénées à l'Oural*. Biotope Editions, Mèze.

Etienne, D., Ruffaldi, P., Dupouey, J.L., Georges-Leroy, M., Ritz, F., Dambrine, E. (2013). Searching for ancient forests: A 2000 years history of land use in northeastern French forests deduced from the pollen compositions of closed depressions. *The Holocene*, 23(5), 678–691.

European Commission (2014). No net loss initiative [Online]. Available at: http://ec.europa.eu/environment/nature/biodiversity/nnl/index_en.htm.

European Commission (2018). Agricultural land abandonment in the EU within 2015–2030. Paper, EC.

European Parliament (2009). European Parliament resolution of 3 February 2009 on Wilderness in Europe (2008/2210(INI)).

Evans, D. and Arvela, M. (2011). Assessment and reporting under article 17 of the habitats directive explanatory, notes and guidelines for the period 2007–2012. Technical report, European Topic Centre on Biological Diversity.

Exavier, R. and Zeilhofer, P. (2020). OpenLand: Software for quantitative analysis and visualization of land use and cover change. *The R Journal*, 12(2), 359.

Eychenne, C. (2018). Le pastoralisme entre mythes et réalités : une nécessaire objectivation – l'exemple des Pyrénées. *Géocarrefour*, 92, 1–15.

Falcucci, A., Maiorano, L., Boitani, L. (2007). Changes in land-use/land-cover patterns in Italy and their implications for biodiversity conservation. *Landscape Ecology*, 22(4), 617–631.

FAO (2020). Global forest resources assessment 2020. Terms and definitions. Technical report, FAO, Forest Resources Assessment (FRA), Rome.

Feller, M., Hooley, D., Dreher, T., East, I., Jung, R. (1979). *Wilderness in Victoria: An Inventory*. Monash Publications in Geography No. 21, Department of Geography, Monash University.

Fernandez-Manjarres, J.F., Ruiz-Benito, P., Zavala, M.A., Camarero, J.J., Pulido, F., Proenca, V., Navarro, L., Sansilvestri, R., Granda, E., Marques, L. *et al.* (2018). Forest adaptation to climate change along steep ecological gradients: The case of the Mediterranean-temperate transition in south-western Europe. *Sustainability*, 10(9), 3065–3076.

Filion, L. and Gärtner, H. (2010). Dendrogéomorphologie. In *La dendroécologie : principes, méthodes et applications*, Payette, S. and Filion, L. (eds). Presses de l'Université Laval, Quebec.

Finney, S.C. and Edwards, L.E. (2016). The "Anthropocene" epoch: Scientific decision or political statement? *GSA Today*, 26, 4–10.

Flanagan, T.S. and Anderson, S. (2008). Mapping perceived wilderness to support protected areas management in the San Juan National Forest, Colorado. *Forest Ecology and Management*, 256, 1039–1048.

FNSAFER (2004). *La fin des paysages. Livre blanc pour une gestion ménagère de nos espaces ruraux*. Ed. Fédération Nationale des SAFER.

Folke, C., Carpenter, S., Walker, B., Scheffer, M., Elmqvist, T., Gunderson, L., Holling, C.S. (2004). Regime shifts, resilience, and biodiversity in ecosystem management. *Annual Review of Ecology Evolution and Systematics*, 35, 557–581.

Fonseca, C.R. (2009). The silent mass extinction of insect herbivores in biodiversity hotspots. *Conservation Biology*, 23, 1507–1515.

Fortibuoni, T., Borme, D., Franceschini, G., Giovanardi, O., Raicevich, S. (2016). Common, rare or extirpated? Shifting baselines for common angelshark, Squatina squatina (Elasmobranchii: Squatinidae), in the Northern Adriatic Sea (Mediterranean Sea). *Hydrobiologia*, 772(1), 247–259.

Fressoz, J.B. (2017). Le prix de l'onction géologique. *La Recherche*, 520, 90.

Fuller, D., Ling, Q., Van Etten, J., Manning, K., Castillo, C., Kingwell-Banham, E., Weisskopf, A., Qin, L., Sato, Y-I., Hijmans, R. (2011). The contribution of rice agriculture and livestock to prehistoric methane levels: An archeological assessment. *Holocene*, 21, 743–759.

Fyfe, R.M., Woodbridge, J., Roberts, N. (2014). From forest to farmland: Pollen inferred land cover changes across Europe using the pseudobiomization approach. *Global Change Biology*, 20, 1197–1212.

Galetti, M. (2004). Parks of the Pleistocene: Recreating the Cerrado and the Pantanal with megafauna. *Natureza & Conservação – Revista Brasileira de Conservação Da Natureza*, 2, 95–101.

Gallet, S., Bioret, F., Sawtschuk, J. (2011). La restauration des végétations des hauts de falaises du littoral atlantique, vers une évaluation globale. *Sciences eaux & territoires*, 5(2), 12–19.

Gallet, S., Jaunatre, R., Regnery, B., Alignan, J.-F., Heckenroth, A., Muller, I., Bernez, I., Combroux, I., Glasser, T., Jund, S. *et al.* (2017). L'écologie de la restauration en France. Dynamique actuelle et rôle d'un réseau multi-acteurs, REVER. *Naturae*, Article 7.

Galop, D. and Bertrand, G.P. (1998). *La forêt, l'homme et le troupeau dans les Pyrénées : 6000 ans d'histoire de l'environnement entre Garonne et Méditerranée*. Éditions Médiriennes, Toulouse.

Galop, D., Houet, T., Mazier, F., Leroux G., Rius, D. (2011). Grazing activities and biodiversity history in the Pyrénées – The use of paleoecology and historical ecology to provide new insights on high-altitude ecosystems in the framework of a Human-Environment Observatory. *PAGES News*, 19(2), 53–56.

Gardner, T.A., Von Hase, A., Brownlie, S., Ekstrom, J.M.M., Pilgrim, J.D., Savy, C.E., Stephens, R.T.T., Treweek, J., Ussher, G.T. *et al.* (2013). Biodiversity offsets and the challenge of achieving no net loss. *Conservation Biology*, 27(6), 1254–1264.

Gaussen, H. (1951). Le dynamisme des biocénoses végétales. *Annales de biologie*, 27(2), 9–22.

Gautier, E. (1994). Interférence des facteurs anthropiques et naturels dans le processus d'incision sur une rivière alpine – l'exemple du Buëch (Alpes du sud). *Revue de géographie de Lyon*, 69(1), 57–62.

Geldmann, J., Joppa, L.N., Burgess, N.D. (2014). Mapping change in human pressure globally on land and within protected areas. *Conservation Biology*, 28, 1604–1616.

Gemenne, F. and Rankovic, A. (2019). *Atlas de l'Anthropocène*. Presses de Sciences Po, Paris.

Génot, J.-C. (2008). *La nature malade de la gestion*. Sang de la Terre, Paris.

Génot, J.-C. (2010). *La nature malade de la gestion : la gestion de la biodiversité ou la domination de la nature*. Sang de la Terre, Paris.

Génot, J.-C. and Schnitzler, A. (2012). *La France des friches*. Éditions Quae, Versailles.

Génot, J.-C. and Schnitzler, A. (2013). Rewilding France via feral nature. *International Journal of Wilderness*, 19(2), 30–33.

GIBOP (2019). Global inventory of biodiversity offset policies (GIBOP). International Union for Conservation of Nature, The Biodiversity Consultancy, Durrell Institute of Conservation & Ecology [Online]. Available at: https://portals.iucn./offsetpolicy/.

GIEC (Groupe d'experts intergouvernemental sur l'évolution du climat) (2013). Rapport d'évaluation 5 (RE5) : les éléments scientifiques (Groupe de travail I) – Résumé à l'intention des décideurs. Report, GIEC.

Giglio, V.J., Luiz, O.J., Gerhardinger, L.C. (2015). Depletion of marine megafauna and shifting baselines among artisanal fishers in eastern Brazil. *Animal Conservation*, 18(4), 348–358.

Gilg, O. (2005). *Old-Growth Forests: Characteristics, Conservation and Monitoring*, vol. 74 bis. ATEN, Montpellier.

Gioria, M., O'Flynn, C., Osborne, B.A. (2018). A review of the impacts of major terrestrial invasive alien plants in Ireland. *Proceedings of the Royal Irish Academy*, 118, 157–179.

Gippoliti, S. and Amori, G. (2006). Ancient introductions of mammals in the Mediterranean Basin and their implications for conservation. *Mammal Review*, 36, 37–48.

Giurgiu, V., Donita, N., Bandiu, C., Radu, S., Cenusa, R., Dissescu, R., Stoiculescu, C., Biris, I. (2001). *Les forêts vierges de Roumanie*. ASBL Foret Wallonne, Brussels.

Glaze, W.H. (2001). The human footprint. *Environmental Science and Technology*, 35, 273.

Glomot, D. (2011). Bocage et Métairies en Haute-Marche au XVe siècle. Aux origines du système d'élevage en prés clos. *Histoire & Sociétés Rurales*, 36, 41–74.

Glowczewski B. and Laurens C. (2018). Le conflit des existences à l'épreuve du climat, ou l'Anthropocène revu par ceux que l'on préfère mettre à la rue ou au musée. In *Penser l'Anthropocène*, Beau, R. and Larrère, C. (eds). Presses de Sciences Po, Paris.

Godet, M. (1977). *Crise de la prévision, essor de la prospective*. Presses Universitaires de France, Paris.

Godet, M. (1986). Introduction to la prospective: Seven key ideas and one scenario method. *Futures*, 18, 134–157.

Godet, L. (2010). La "nature ordinaire" dans le monde occidental. *L'Espace géographique*, 39(4), 295–308.

Godet, L. (2017). Les oiseaux anthropophiles : définition, typologie et conservation. *Annales de Géographie*, 716, 492–517.

Godet, L. and Devictor, V. (2018). What conservation does. *Trends in Ecology and Evolution*, 33, 720–730.

Godoy, N., Gelcich, S., Vásquez, J.A., Castilla, J.C. (2010). Spearfishing to depletion: Evidence from temperate reef fishes in Chile. *Ecological Applications*, 20(6), 1504–1511.

Goeldner-Gianella, L. (2007). Dépoldériser en Europe occidentale. *Annales de Géographie*, 4(656), 339–360.

Goeldner-Gianella, L. (2012). Changement climatique et dépoldérisation : le rôle des acteurs et le poids des représentations sociales sur les côtes d'Europe atlantique. *Quaderni*, 71 [Online]. Available at: http://journals.openedition.org/quaderni/527 [Accessed 4 February 2021].

Goeldner-Gianella, L. and Verger, F. (2009). Du "polder" à la "dépoldérisation" ? *L'Espace géographique*, 4(38), 376–377.

Gonçalves, B., Marques, A., Soares, A.M., Pereira, H.M. (2015). Biodiversity offsets: From current challenges to harmonized metrics. *Current Opinion in Environmental Sustainability*, 14, 61–67.

Gordon, A., Bull, J.W., Wilcox, C., Maron, M. (2015). Perverse incentives risk undermining biodiversity offset policies. *Journal of Applied Ecology*, 52, 532–537.

Gourmelon, F., Houet, T., Voiron, C., Joliveau T. (2012). La géoprospective. Points de vue conceptuels et applicatifs de l'apport du spatial à la prospective. *L'Espace géographique*, 2, 97–98.

Govedar, Z., Krstic, M., Keren, S., Babic, V., Zlokapa, B., Kanjevac, B. (2018). Actual and balanced stand structure: Examples from beech-fir-spruce old-growth forests in the area of the dinarides in bosnia and herzegovina. *Sustainability*, 10, 540–551.

Gradstein, F.M., Ogg, J.G., Schmitz, M.D., Ogg, G.B. (2012). *The Geologic Time Scale 2012*. Elsevier, Boston.

Gramond, D. (2014). Géohistoire environnementale : contours sémantiques et conceptuels. Discussions sur les héritages et patrimoines reconnus aux zones humides fluviales. *Développement durable et territoires*, 5, 1–11.

Grayson, D.K. and Meltzer, D.J. (2003). A requiem for North American overkill. *Journal of Archaeological Science*, 30, 585–593.

Gregory, R., Lichtenstein, S., MacGregor, D. (1993). The role of past states in determining reference points for policy decisions. *Organizational Behavior and Human Decision Processes*, 55, 195–206.

Greslier, N. (1993). Inventaire des forêts subnaturelles de l'arc alpin français. End of study Thesis, FIF, Nancy.

Guay, L., Wyatt, S., Hebert, M., Desbiens, C. (2013). L'aménagement écosystémique des forêts au Québec : questions et pistes de réflexion autour de la prise en compte des enjeux sociaux. Report, Ministère des forêts, Quebec.

Guénette, S. and Gascuel, D. (2012). Shifting baselines in European fisheries: The case of the Celtic Sea and Bay of Biscay. *Ocean & Coastal Management*, 70, 10–21.

Guerrero-Gatica, M., Aliste, E., Simonetti, J.A. (2019). Shifting gears for the use of the shifting baseline syndrome in ecological restoration. *Sustainability*, 11(5), 1458.

Guetté, A., Carruthers-Jones, J., Godet, L., Robin M. (2018). "Naturalité" : concepts et méthodes appliqués à la conservation de la nature. *Cybergeo: European Journal of Geography* [Online]. Available at: https://journals.openedition.org/cybergeo/29140#toc.

Guillet, F. and Semal, L. (2018). Policy flaws of biodiversity offsetting as a conservation strategy. *Biological Conservation*, 221, 86–90.

Guinier, E. (1905). La forêt vierge de Doussard et la forêt du crêt du Maure. *Revue des eaux et forêts*, 577–590.

Guinier, P. (1950). Foresterie et protection de la nature : l'exemple de Fontainebleau. *Revue forestière française*, 12, 703–717.

Guyon, L.J. and Battaglia, L.L. (2018). Ecological characteristics of floodplain forest reference sites in the Upper Mississippi River System. *Forest Ecology and Management*, 427, 208–216.

Haguenauer, C. (1991). Comprendre par les cycles... et les cycles pour apprendre ou le concept de cycle, indicateur de la connaissance, des sciences de la nature à l'écologie forestière. PhD Thesis, Université Henri Poincaré-Nancy 1, Vandoeuvre-les-Nancy.

Hall, M. (2005). *Earth Repair: A Transatlantic History of Environmental Restoration*. University of Virginia Press, Charlottesville.

Halpern, B.S., Walbridge, S., Selkoe, K.A., Kappel, C.V., Micheli, F., D'Agrosa, C., Bruno, J.F., Casey, K.S., Ebert, C., Fox, H.E. *et al.* (2008). A global map of human impact on marine ecosystems. *Science*, 319, 948–952.

Hannah, L., Lohse, D., Hutchinson, C., Carr, J.L., Lankerani, A. (1994). A preliminary inventory of human disturbance of world ecosystems. *Ambio*, 23, 246–250.

Hannah, L., Carr, J.L., Lankerani, A. (1995). Human disturbance and natural habitat: A biome level analysis of a global data set. *Biodiversity and Conservation*, 4, 128–155.

Harari, Y.N. (2015). *Sapiens: A Brief History of Humankind*. Vintage, London.

Haraway, D. (2015). Anthropocene, Capitalocene, Plantationocene, Chthulucene: Making Kin. *Environmental Humanities*, 6, 159–165.

Harmon, M., Franklin, J., Swanson, F., Sollins, P., Gregory, S., Lattin, J., Anderson, N., Cline, S., Aumen, N., Sedell, J. *et al.* (1986). Ecology of coarse woody debris in temperate ecosystems. *Advances in Ecological Research*, 15, 133–302.

Hatem, F. (1993). *La prospective : pratiques et méthodes*. Economica, Paris.

Hatzfeld, H. (2009). Les enjeux du paysage. In *Le paysage, un projet politique*, Bédard, M. (ed.). Presses de l'Université du Québec, Quebec.

Hawkins, C.P., Olson, J.R., Hill, R.A. (2010). The reference condition: Predicting benchmarks for ecological and water-quality assessments. *Journal of the North American Benthological Society*, 29, 312–343.

Hédin, L. (1951). Les conditions d'exploitation de la prairie en Normandie, depuis le XIXe siècle. *Annales de Normandie*, 1, 45–69.

Heinselman, M.L. (1981). Fire and succession in the conifer forests of northern North America. In *Forest Succession: Concepts and Application*, West, D.C., Shugart, H.H., Botkin, D.B. (eds). Springer-Verlag, New York.

Helman, P.H., Jones, A.D., Pigram, J.J.J., Smith, J.M.B. (1976). *Wilderness in Australia. Eastern New South Wales and South East Queensland*. Department of Geography, University of New England, Armidale.

Herlihy, A.T., Paulsen, S.G., Sickle, J.V., Stoddard, J.L., Hawkins, C.P., Yuan, L.L. (2008). Striving for consistency in a national assessment: The challenges of applying a reference-condition approach at a continental scale. *Journal of the North American Benthological Society*, 27, 860–877.

Hervieu, B. and Viard, J. (1996). *Au bonheur des campagnes*. Ed de l'Aube, Marseille.

Higgs, E. (1997). What is good eccological restoration? *Conservation Biology*, 11(2), 338–348.

Higgs, E., Falk, D., Guerrini, A., Hall, M., Harris, J., Hobbs, R., Jackson, S., Rhemtulla, J., Throop, W. (2014). The changing role of history in restoration ecology. *Frontiers in Ecology and Environment*, 12(9), 499–506.

Hilding-Rydevik, T., Moen, J., Green, C. (2018). Baselines and the shifting baseline syndrome: Exploring frames of reference in nature conservation. In *Issues and Concepts in Historical Ecology: The Past and Future of Landscapes and Regions*, Crumley, C.L., Lennartsson, T., Westin, A. (eds). Cambridge University Press, Cambridge.

Hobbs, R.J., Arico, S., Aronson, J., Baron, J.S., Bridgewater, P., Cramer, V.A., Epstein, P.R., Ewel, J.J., Klink, C.A., Lugo, A.E. *et al.* (2006). Novel ecosystems: Theoretical and management aspects of the new ecological world order. *Global Ecology and Biogeography*, 15(1), 1–7.

Hobbs, R.J., Higgs, E.S., Hall, C.M. (2013). *Novel Ecosystems: Intervening in the New Ecological World Order.* Wiley-Blackwell, Chichester.

Hofreiter, M. and Barnes, I. (2010). Diversity lost: Are all Holarctic large mammal species just relict populations? *BMC Biology*, 8, 46.

Holling, C.S. (1992). Cross-scale morphology, geometry, and dynamics of ecosystems *Ecological Monographs*, 62, 447–502.

Holling, C.S. (2001). Understanding the complexity of economic, ecological, and social systems. *Ecosystems*, 4(5), 390–405.

Holling, C.S., Gunderson, L.H., Peterson, G.D. (2002). Sustainability and panarchies. In *Panarchy, Understanding Transformations in Human and Natural Systems*, Gunderson, L.H. and Holling, C.S. (eds). Island Press, Washington, London.

Hornborg, A. (2015). The political ecology of the technocene: Uncovering ecologically unequal exchange in the world-system. In *The Anthropocene and the Global Environmental Crisis: Rethinking Modernity in a New Epoch*, Hamilton, C., Bonneuil, C., Gemenne, F. (eds). Routledge, London.

Hosner, D., Wagner, M., Tarasov, P.E., Chen, X., Leipe, C. (2016). Spatiotemporal distribution patterns of archaeological sites in China during the Neolithic and Bronze Age: An overview. *Holocene*, 26, 1576–1583.

Hou, W., Zhai, L., Qiao, Q., Walz, U. (2019). Monitoring the intensity of human impacts on anthropogenic landscape: A mapping case study in Beijing, China. *Ecological Indicators*, 102, 382–393.

Houet, T. (2006). Occupation des sols et gestion de l'eau : modélisation prospective en paysage agricole fragmenté (Application au SAGE du Blavet). PhD Thesis, Université Rennes 2 [Online]. Available at: https://tel.archives-ouvertes.fr/tel-00389835.

Houet, T. (2015). Usages de modèles spatiaux pour la prospective. *Revue internationale de géomatique*, 25(1), 123–143.

Houet, T., Loveland, T.R., Hubert-Moy, L., Napton, D., Gaucherel, C., Barnes C. (2010). Exploring subtle land use and land cover changes: A framework based on future landscape studies. *Landscape Ecology*, 25(2), 249–266.

Houet, T., Hubert-Moy, L., Tyssot, C. (2011). Fine scale spatialised prospective modelling – A methodological approach. Application to water management in Brittany. *Revue internationale de géomatique*, 21, 67–93.

Houet, T., Ribière, O., Vacquié, L., Vidal, F., Galop D. (2012). Analyse spatiale de l'évolution des modes d'occupation et d'usages des sols sur le Vicdessos de 1942 à nos jours. *Sud-Ouest Européen*, 33, 41–56.

Houet, T., Vacquié, L., Sheeren, D. (2015). Evaluating the spatial uncertainty of future land abandonment in a mountainous valley (Vicdessos, Pyrenees – France): Insights from model parameterization and experiments. *Journal of Mountain Science*, 12(5), 1095–1112.

Houet, T., Marchadier, C., Bretagne, G., Moine, M.P., Aguejdad, R., Viguié, V., Bonhomme, M., Lemonsu, A., Avner, P., Hidalgo, J. *et al.* (2016a). Combining narratives and modeling approaches to simulate fine scale and long-term urban growth scenarios for climate adaptation. *Environmental Modelling and Software*, 86, 1–13.

Houet, T., Aguejdad, R., Doukari, O., Battaia, G., Clarke, K. (2016b). Description and validation of a "non path-dependent" model for projecting contrasting urban growth futures. *Cybergeo*, 759 [Online]. Available at: http://cybergeo.revues.org/27397.

Houet, T., Grémont, M., Vacquié, L., Forget, Y., Bourrier, F., Peltier, A., Puissant, A., Bernardie, S., Grandjean, G. (2017). Downscaling scenarios of future land use and land cover changes using a participatory approach: An application to mountain risk assessment in the Pyrenees (France). *Regional Environmental Change*, 17(8), 2293–2307.

Hough, P. and Robertson, M. (2009). Mitigation under Section 404 of the Clean Water Act: Where it comes from, what it means. *U.S. Environmental Protection Agency Papers*, 269.

Houghton, R.A. and Hackler, J.L. (2003). Sources and sinks of carbon from land-use change in China. *Global Biogeochemical Cycles*, 17, 1034.

Houzard, G. (1986). Dégradations et restauration en Ecouves de 1666 à nos jours. *Hommes et terres du nord*, 2(1), 227–230.

Houzard, G. and Lecointe, A. (1991). Etude biogéographique des friches. Premiers résultats / A biogeographical study of waste lands: Preliminary results. *Revue de géographie de Lyon*, 66(1), 38–46.

Hrabanski, M. (2015). The biodiversity offsets as market-based instruments in global governance: Origins, success and controversies. *Ecosystem Services*, 1–9.

Huggins, D.G. and Dzialowski, A. (2005). Identification and quantification of reference conditions associated with lotic ecosystems of the mid-continent region of the US: A summary of approaches and factors applicable to Heartland Network facilities. Report, Central Plains Center for BioAssessment Kansas Biological Survey Lawrence.

Hughes, K.A., Fretwell, P., Rae, J., Holmes, K., Fleming, A. (2011). Untouched Antarctica: Mapping a finite and diminishing environmental resource. *Antarctic Science*, 23, 537–548.

Hurand, B. (2014). Irréductible nature. In *Y a-t-il du sacré dans la nature ?* Hurand, B. and Larrère, C. (eds). Publications de la Sorbonne, Paris.

INPN (2020). La biodiversité en France – 100 chiffres expliqués sur les espaces protégés. UMS PatriNat (OFB-CNRS-MNHN), Paris.

International Finance Corporation (IFC) (2012). Performance standards [Online]. Available at: https://www.ifc.org/wps/wcm/connect/Topics_Ext_Content/IFC_External_Corporate_Site/Sustainability-At-IFC/Policies-Standards/Performance-Standards.

International Union for Conservation of Nature (IUCN) (2016). IUCN policy on biodiversity offsets [Online]. Available at: https://www.iucn.org/theme/business-and-biodiversity/our-work/business-approaches-and-tools/biodiversity-offsets.

IPBES (2019). Summary for policymakers of the global assessment report on biodiversity and ecosystem services of the Intergovernmental Science-Policy Platform on Biodiversity and Ecosystem Services. Report, IPBES Secretariat, Bonn.

Isaac, N.J., Turvey, S.T., Collen, B., Waterman, C., Baillie, J.E. (2007). Mammals on the EDGE: Conservation priorities based on threat and phylogeny. *PloS One*, 2, e296.

Jacob, C., Vaissière, A.C., Bas, A., Calvet, C. (2016). Investigating the inclusion of ecosystem services in biodiversity offsetting. *Ecosystem Services*, 21, 92–102.

Jansen, P., van der Stoep, J., Keulartz, J., Jochemsen, H. (2016). In search of religious elements in the Dutch nature policy. *Journal for the Study of Religion, Nature and Culture*, 10(1), 91–116.

Jansen, P., van der Stoep, J., Keulartz, J., Jochemsen, H. (2017). Wistful wilderness: Communication about "new" nature in the Netherlands. *Journal of Environmental Policy and Planning*, 19(2), 197–213.

Janzen, D. (1998). Gardenification of wildland nature and the human footprint. *Science*, 279, 1312v1313.

Jaroszewicz, B., Cholewińska, O., Gutowski, J.M., Samojlik, T., Zimny, M., Latałowa, M. (2019). Białowieża forest – A relic of the high naturalness of European forests. *Forests*, 10, 849 [Online]. Available at: https://doi.org/10.3390/f10100849.

Jepson, P. (2016). A rewilding agenda for Europe: Creating a network of experimental reserves. *Ecography*, 39(2), 117–124.

Jepson, P. and Ladle, R. (2010). *Conservation. A Beginner's Guide*. Oneworld Publications, Oxford.

Jones, E. (1945). The structure and reproduction of the virgin forest of the north temperate zone. *New Phytologist*, 44(2), 130–148.

Jones, C.G., Lawton, J.H., Shachak, M. (1994). Organisms as ecosystem engineers. *Oikos*, 69, 373–386.

Jørgensen, D. (2015). Rethinking rewilding. *Geoforum*, 10.1016/j.geoforum.2014.11.016.

Jouffroy-Bapicot, I., Vannière, B., Gauthier, É., Richard, H., Monna, F., Petit, C. (2013). 7000 years of vegetation history and land-use changes in the Morvan mountains (France): A regional synthesis. *The Holocene*, 23(12), 1888–1902.

de Jouvenel, B. (1964). *L'art de la conjecture*. Editions du Rocher, Monaco.

Kai, Z., Woan, T.S., Jie, L., Goodale, E., Kitajima, K., Bagchi, R., Harrison, R.D. (2014). Shifting baselines on a tropical forest frontier: Extirpations drive declines in local ecological knowledge. *PLoS One*, 9(1), e86598.

Kaplan, S. (1987). Aesthetics, affect and cognition: Environmental preference from an evolutionary perspective. *Environment and Behavior*, 19(1), 3–32.

Kaplan, J.O., Krumhardt, K.M., Zimmermann, N. (2009). The prehistoric and preindustrial deforestation of Europe. *Quaternary Science Reviews*, 28(27), 3016–3034.

Katikiro, R.E. (2014). Perceptions on the shifting baseline among coastal fishers of Tanga, Northeast Tanzania. *Ocean & Coastal Management*, 91, 23–31.

Keenleyside, C., Tucker, G., McConville, A. (2010). *Farmland Abandonment in the EU: An Assessment of Trends and Prospects*. Institute for European Environmental Policy, London.

Kéfi, S. (2012). Des écosystèmes sur le fil : comment certains écosystèmes basculent d'un état à un autre. *Regards de la Société d'Ecologie et d'Evolution*, 37 [Online]. Available at: https://www.sfecologie.org/regard/r37-hysteresis-sonia-kefi/.

Keilty, K., Beckley, T.M., Sherren, K. (2016). Baselines of acceptability and generational change on the Mactaquac hydroelectric dam headpond (New Brunswick, Canada). *Geoforum*, 75, 234–248.

Kershner, J.L., Roper, B.B., Bouwes, N., Henderson, R., Archer, E. (2004). An analysis of stream habitat conditions in reference and managed watersheds on some federal lands within the Columbia River Basin. *North American Journal of Fisheries Management*, 24, 1363–1375.

Kirkpatrick, J.B. (1979). Hydro-electric development and wilderness. Report to the department of environment. Assessment of the HEC Report on the Lower Gordon Rover Development Stage Two. Department of the Environment, Hobart.

Klatzmann, J. (1955). *La localisation des cultures et des productions animales en France*. INSEE, Paris.

Kliskey, A.D. (1994). A comparative analysis of approaches to wilderness perception mapping. *Journal of Environmental Management*, 41, 199–236.

Kliskey, A.D. (1998). Linking the wilderness perception mapping concept to the recreation opportunity spectrum. *Environmental Management*, 22, 79–88.

Kliskey, A.D. and Kearsely, G.W. (1993). Mapping multiple perceptions of wilderness in southern New Zealand. *Applied Geography*, 13, 203–223.

Kliskey, A.D. and Kearsely, G.W. (1994). Mapping multiple perceptions of wilderness in southern New Zealand II: An alternative multivariate approach. *Applied Geography*, 14, 308–326.

Kolbert, E. (2014). *The Sixth Extinction. An Unnatural History*. Henry Holt and Company, New York.

Kondolf, G.M. (2006). River restoration and meanders. *Ecology and Society*, 11 [Online]. Available at: https://doi.org/10.5751/ES-01795-110242.

Koop, H. and Hilgen, P. (1987). Forest dynamics and regeneration mosaic shifts in unexploited beech (*Fagus sylvatica*) stands at Fontainebleau (France). *Forest Ecology and Management*, 20, 135–150.

Kopf, R.K., Finlayson, C.M., Humphries, P., Sims, N.C., Hladyz, S. (2015). Anthropocene baselines: Change and managing biodiversity in human-dominated aquatic ecosystems. *BioScience*, 65(8), 798–811.

Kopnina, H., Leadbeater, S., Cryer, P. (2019). Learning to rewild: Examining the failed case of the Dutch "New Wilderness" Oostvaardersplassen. *International Journal of Wilderness*, 25(3), 72–88.

Korpel, S. (1995). *Die Urwälder der Westkarpaten*. Gustav Fischer, Stuttgart.

Kraszewski, M. (2019). Évaluation du taux d'artificialisation en France : comparaison des sources Teruti-Lucas et fichiers fonciers. Commissariat général au développement durable [Online]. Available at: https://www.statistiques.developpement-durable.gouv.fr/sites/default/files/2019-08/datalab-56-evaluation-du-taux-d-artificialisation-en-france-aout2019.pdf.

Landres, P. (2010). Let it be: A hands-off approach to preserving wildness in protected areas. In *Beyond Naturalness: Rethinking Park and Wilderness Stewardship in an Era of Rapid Change*, Cole, D.N. and Yung, L. (eds). Island Press, Washington, DC.

Landres, P., Brunson, M.W., Merigliano, L., Sydoriak, C., Morton, S. (2000). Naturalness and wildness: The dilemma and irony of managing wilderness. *Proceedings RMRS-P-15*, 5, 377–381.

Larsen, J.A. (1980). *The Boreal Ecosystem*. Academic Press, New York.

Lasanta, T., Arnáez, J., Pascual, N., Ruiz-Flaño, P., Errea, M.P., Lana-Renault, N. (2017). Space–time process and drivers of land abandonment in Europe. *CATENA*, 149, 810–823.

Le Floc'h, E. and Aronson, J. (1995). Écologie de la restauration. Définition de quelques concepts de base. *Natures sciences sociétés*, 3, special edition "Recréer la nature", 29–35.

Le Floch, S., Devanne, A.S., Deffontaines, J.P. (2005). La "fermeture du paysage" : au-delà du phénomène, petite chronique d'une construction sociale. *L'Espace geographique*, 34(1), 49–64.

Leakey, R. and Lewin, R. (1996). *The Sixth Extinction. Patterns of Life and the Future of Humankind*. Anchor, New York.

Lebeau, R. (1969). *Les grands types de structures agraires dans le monde*. Masson, Paris.

Leeney, R.H. and Downing, N. (2016). Sawfishes in The Gambia and Senegal – Shifting baselines over 40 years. *Aquatic Conservation: Marine and Freshwater Ecosystems*, 26(2), 265–278.

Lemée, G. (1989). Structure et dynamique de la hêtraie des réserves biologiques de la forêt de Fontainebleau : un cas de complexe climacique de forêt feuillue monospécifique tempérée. *Acta Oecologica – Oecologia Generalis*, 10(2), 155–174.

Lemoine, G. (2005). Gestion et plans de gestion pour les milieux naturels : paradoxe ou utopie ? *La Garance voyageuse*, 70, 24–29.

Leopold, A. (1949). *A Sand County Almanac*. Oxford University Press, Oxford.

Leroy, C. (2006). Le degré de naturalité des forêts de la réserve naturelle des rochers et tourbières du pays de Bitche. Détermination d'indicateurs de naturalité, définition d'une méthodologie de suivi du degré de naturalité et établissement de l'état initial de certains sites. End of study thesis, FIF, Nancy.

Lespez, L. (2012). Les temps de l'environnement et des paysages des systèmes fluviaux au cours de l'Holocène. Normandie, Grèce, Mali. HDR, Université Caen Basse Normandie.

Lespez, L. and Dufour, S. (2021). Les hybrides, la géographie de la nature et de l'environnement. *Annales de Géographie*, 737, 58–85.

Lespez, L. and Germaine, M.A. (2016). La rivière désaménagée ? Les paysages fluviaux et l'effacement des seuils et des barrages en Europe de l'Ouest et en Amérique du Nord-Est. *Bulletin de la Société Géographique de Liège*, 67, 223–254.

Lesslie, R.G. and Mackey, B.G. (1988). A computer-based method of wilderness evaluation. *Environmental Conservation*, 15, 225–232.

Lesslie, R.G. and Taylor, S.G. (1983). Wilderness in South Australia. Occasional Paper No. 1, Centre for Environmental Studies, University of Adelaide.

Lesslie, R.G. and Taylor, S.G. (1985). The wilderness continuum concept and its implications for Australian wilderness preservation policy. *Biological Conservation*, 32, 309–333.

Lesslie, R.G., Mackey, B.G., Shulmeister, J. (1988). *Wilderness Quality in Tasmania*. Australian Heritage Commission, Canberra.

Levrel, H., Guillet, F., Lombard-Latune, J., Delforge, P., Frascaria-Lacoste, N. (2018). Application de la séquence éviter-réduire-compenser en France : le principe d'additionnalité mis à mal par 5 dérives. *VertigO*, 18(2).

Lévy, J. and Lussault, M. (2013). *Dictionnaire de géographie et de l'espace des sociétés*. La Documentation Française, Paris.

Lewis, S.L. and Maslin, M.A. (2015). Defining the Anthropocene. *Nature*, 519, 171–180.

Lewis, S.L. and Maslin, M.A. (2018). L'an 1610 de notre ère. In *Penser l'Anthropocène*, Beau, R. and Larrère, C. (eds). Presses de Sciences Po, Paris.

Li, X., Dodson, J., Zhou, J., Zhou, X. (2008). Increases of population and expansion of rice agriculture in Asia, and anthropogenic methane emissions since 5000 BP. *Quaternary International*, 202, 41–50.

Liébault, F., Cassel, M., Jantzi, H., Tacon, S., Talaska, N. (2010). Régime sédimentaire et morphohologie des rivières en tresses alpines et méditerranéennes. Technical report, Cemagref, Grenoble.

Lied, K. and Bakkehøi, S. (1980). Empirical calculation of snow-avalanche run-out distance based on topographic parameters. *Journal of Glaciology*, 26(94), 165–177.

Loinger, G. (2004). *La prospective régionale, de chemins en desseins*. Éditions de l'Aube, La Tour d'Aigues, Paris.

Lorimer, J. (2015). *Wildlife in the Anthropocene. Conservation after Nature*. University of Minnesota Press, Minnesota, London.

Lovelock J. (1979). *Gaia, a New Look at Life on Earth*. Oxford University Press, Oxford.

Lowenhaupt Tsing, A. (2017). *Le champignon de la fin du monde*. La Découverte, Paris.

Lowenthal, D. (1997). European landscape transformations: The rural residue. In *Understanding Ordinary Landscapes*, Groth, P. and Bressi, W. (eds). Yale University Press, New Haven.

Lucas, R.C. (1964). Wilderness perception and use: The example of the Boundary Waters Canoe Area. *Natural Resources Journal*, 3, 394–411.

Lucas, S.G. (2018). The GSSP method of chronostratigraphy: A critical review. *Frontiers in Earth Science*, 6, 191.

Luginbühl, Y. and Terrasson, D. (2013). *Paysage et développement durable*. Éditions Quae, Versailles.

Lunde, K.B., Cover, M.R., Mazor, R.D., Sommers, C.A., Resh, V.H. (2013). Identifying reference conditions and quantifying biological variability within benthic macroinvertebrate communities in perennial and non-perennial Northern California streams. *Environmental Management*, 51, 1262–1273.

MacDonald, D., Crabtree, J.R., Wiesinger, G., Dax, T., Stamou, N., Fleury, P., Gutierrez Lazpita, J., Gibon, A. (2000). Agricultural abandonment in mountain areas of Europe: Environmental consequences and policy response. *Journal of Environmental Management*, 59(1), 47–69.

Machado, A. (2004). An index of naturalness. *Journal for Nature Conservation*, 12, 95–110.

Maciejewski, L., Lepareur, F., Viry, D., Bensettiti, F., Puissauve, R., Touroult, J. (2016). État de conservation des habitats : propositions de définitions et de concepts pour l'évaluation à l'échelle d'un site Natura 2000. *Revue d'écologie (La Terre et la Vie)*, 71, 3–20.

MacPhee, R.D. and Marx, P.A. (1997). The 40,000-year plague: Humans, hyperdisease, and first-contact extinctions. In *Natural Change and Human Impact in Madagascar*, Goodman, S.M. and Patterson, B.D. (eds). Smithsonian Institution Press, Washington, DC.

Magny, M. (2021). *L'Anthropocène*. PUF, Paris.

Malm, A. (2016). *Fossil Capital: The Rise of Steam Power and the Roots of Global Warming*. Verso, Brooklyn.

Malme A. (2017). *L'anthropocène contre l'histoire : le réchauffement climatique à l'ère du capital*. La Fabrique Editions, Paris.

Marage, D. (2018). La forêt dans tous ses états : de la conservation à l'évaluation territoriale. HDR, Université Paris 1-Panthéon-Sorbonne, Paris, Washington, DC.

Marage, D. and Lempérière, G. (2005). The management of snags: A comparison in managed and unmanaged ancient forests of the southern French Alps. *Annals of Forest Science*, 62(2), 135–142.

Marage, D., Rameau, J.C., Garraud, L. (2006). Soil seed banks and vegetal succession in the Southern Alps: Effects of historic and ecologic factors. *Canadian Journal of Botany*, 84, 99–111.

Marage, D., Lemperière, G., Petit-Berghem, Y., Simon, L. (in press). Mieux connaître est ce mieux gérer ? Recours à la casuistique socio-écologique. In *De la réserve intégrale à la nature ordinaire*, Luglia, R., Beau, R., Treillard, A. (eds). Presses universitaires de Rennes, Paris.

Margules, C. and Pressey, R. (2000). Systematic conservation planning. *Nature*, 405, 243–253.

Maris, V. (2014). *Nature à vendre. La limite des services écosystémiques*. Éditions Quae, Versailles.

Maris, V. (2018). *La part sauvage du Monde. Penser la nature dans l'Anthropocène*. Le Seuil, Paris.

Maron, M., Bull, J.W., Evans, M.C., Gordon, A. (2015). Locking in loss: Baselines of decline in Australian biodiversity offset policies. *Biological Conservation*, 192, 504–512.

Maron, M., Brownlie, S., Bull, J.W., Evans, M.C., von Hase, A., Quétier, F., Watson, J.E.M., Gordon, A. (2018). The many meanings of no net loss in environmental policy. *Nature Sustainability*, 1, 19–27.

Marris, E. (2009). Reflecting the past. *Nature*, 462, 30–32.

Marris, E. (2011). *Rambunctious Garden. Saving the Nature in a Post-Wild*. Bloomsbury, New York.

Marris, E., Mascaro, J., Ellis, E.C. (2013). Perspective: Is everything a novel ecosystem? If so, do we need the concept? In *Novel Ecosystems*, Hobbs, R.J., Higgs, E.S., Hall, C.M. (eds). John Wiley & Sons, Ltd, Chichester.

Martin, P.S. (1967). Prehistoric overkill. In *Pleistocene Extinctions, the Search for the Cause*, Martin, P.S. and Wright, H.E. (eds). Yale University Press, New Haven, London.

Martin, P.S. (1984). Prehistoric overkill: The global model. In *Quaternary Extinctions*, Martin, P.S. and Klein R.G. (eds). University of Arizona Press, Arizona.

Martin, P.S. and Burney, D. (1999). Bring back the elephants! *Wild Earth*, 9, 57–65.

Martin, P.S. and Steadman, D.W. (1999). Prehistoric extinctions on islands and continents. In *Extinctions in Near Time: Causes, Contexts and Consequences*, MacPhee, R.D.E. (ed.). Kluwer Academic/Plenum, New York.

Marty, P. (2005). Activités humaines et production des espaces naturels. In *Les biodiversités : objets, théories, pratiques*, Marty, P., Vivien, F.D., Lepart, J., Larrère, R. (eds). CNRS Editions, Paris [Online]. Available at: https://halshs.archives-ouvertes.fr/halshs-01273833.

Marty, P. and Lepart, J. (2001). Forêts et milieux ouverts : anciennes et nouvelles légitimités. *Bulletin de l'Association de géographes français*, 78(2), 177–189.

Mather, A.S., Fairbairn, J., Needle, C.L. (1999). The course and drivers of the forest transition: The case of France. *Journal of Rural Studies*, 15(1), 65–90.

Mathevet, R. (2004). *Camargue incertaine : sciences, usages et natures*. Buchet Chastel, Paris.

Mathevet, R. and Béchet, A. (2020). *Politiques du flamant rose, vers une écologie du sauvage*. Wildprojets, Marseille.

Mathevet, R., Peluso, N.L., Couespel, A., Robbins, P. (2015). Using historical political ecology to understand the present: Water, reeds, and biodiversity in the Camargue Biosphere Reserve, southern France. *Ecology and Society*, 20(4), 17.

McCallum, M.L. (2007). Amphibian decline or extinction? Current declines dwarf background extinction rate. *Journal of Herpetology*, 41, 483–491.

McCarthy, D.P. (2013). Lichenometry. In *Encyclopedia of Quaternary Science*, Elias, S.A. and Mock, C.J. (eds). Elsevier, Amsterdam.

McCauley, D.J., Hardesty-Moore, M., Halpern, B.S., Young, H.S. (2016). A mammoth undertaking: Harnessing insight from functional ecology to shape de-extinction priority setting. *Functional Ecology*, 31, 1003–1011.

McClenachan, L. (2009). Documenting loss of large trophy fish from the Florida Keys with historical photographs. *Conservation Biology*, 23(3), 636–643.

McCloskey J.M. and Spalding, H. (1989). A reconnaissance-level inventory of the amount of wilderness remaining in the world. *Ambio*, 18, 221–227.

McClung, D. and Lied, K. (1987). Statistical and geometric definition of snow avalanche runout. *Cold Regions Science and Technology*, 13, 107–119.

McClung, D. and Mears, A.I. (1995). Dry flowing avalanche run-up and run-out. *Journal of Glaciology*, 41(138), 359–372.

McClung, D., Mears, A.I., Schaerer, P. (1989). Extreme avalanche runout: Data from four mountain ranges. *Annals of Glaciology*, 13, 180–184.

McKenney, B.A. and Kiesecker, J.M. (2010). Policy development for biodiversity offsets: A review of offset frameworks. *Environmental Management*, 45, 165–176.

McLean, M., Cuetos-Bueno, J., Nedlic, O., Luckymiss, M., Houk, P. (2016). Local stressors, resilience, and shifting baselines on coral reefs. *PloS One*, 11(11).

McNeill, J.R. and Engelke, P. (2016). *The Great Acceleration: An Environmental History of the Anthropocene since 1945*. Harvard University Press, Cambridge.

McNellie, M., Oliver, I., Dorrough, J., Ferriers, S., Newell, G., Gibbons, P. (2020). Reference state and benchmark concepts for better conservation in contemporary ecosystem. *Global Change Biology*, 26, 6702–6714.

Meachen, J. and Samuels, J. (2012). Evolution in coyotes (*Canis latrans*) in response to the megafaunal extinctions. *Proceedings of the National Academy of Sciences*, 109, 4191–4196.

Meadows, D.H., Meadows, D.L., Randers, J., Behrens, I.W.W. (1972). Rapports sur les limites à la croissance. In *Halte à la croissance ?*, Club de Rome (ed.). Fayard, Paris.

Méchin, A. (2020). Dimensionner les mesures de compensation écologique : des outils opérationnels pour une meilleure appropriation par les acteurs de l'aménagement du territoire. Thèse de doctorat, Université Paul-Valéry, Montpellier.

Mendras, H. (1967). *La fin des paysans, innovations et changements dans l'agriculture française*. Sedeis, Paris.

Mendras, H. and Jollivet, M. (1965). Les sociétés rurales françaises : inventaire typologique et étude des changements sociaux. *Economie rurale*, 64, 9–18.

Mengin, N., Bougon, N., Chandesris, A., Oraison, F., Souchon, Y., Valette, L., Villeneuve, B. (2009). Réseau de référence des eaux douces de surface – cours d'eau. ONEMA – CEMAGREF, Lyon.

Meyer, S. (2006). *The End of the Wild*. MIT Press, Cambridge.

Ministère de l'Agriculture (1892). Statistique agricole de la France : résultats généraux de l'enquête décennale de 1892. Imprimerie Nationale, Paris.

Ministère de l'Agriculture (1936). Statistique agricole de la France : résultats généraux de l'enquête de 1929. Imprimerie Nationale, Paris.

Ministère de la transition écologique (2021). Stratégie nationale pour les aires protégées 2030. Report, Ministère de la transition écologique.

Mitchell, F.J.G. (2005). How open were European primeval forests? Hypothesis testing using palaeoecological data. *Journal of Ecology*, 93, 168–177.

Molau, U. and Mølgaard, P. (1996). *ITEX Manual – International Tundra Experiment*. Danish Polar Center, Copenhagen.

Monbiot, G. (2014). *Feral. Rewilding the Land, Sea and Human Life*. Penguin Books, London.

Monnier, L. and Potier, C. (2018). Synthèse Mai 2018 – Novembre 2018 : Inventaire des Inventaires de Friches par Centrale Lille Projets. Laboratoire d'Initiatives Foncières et Territoriales.

Monod, T. (1992). *Plongées profondes. Bathyfolages*. Actes Sud, Arles.

Montembault, D. (2007a). La valorisation des paysages de vallées en Anjou : une campagne plus vraie que nature ? In *Agriculteurs et territoires. Entre productivisme et exigences territoriales*, Madeline, P., Pierre, G., Le Caro, Y. (eds). PUR, Rennes.

Montembault, D. (2007b). Origine, évolution et ré-interprétation des haies de frênes têtards dans les vallées angevines : vers la construction de nouvelles formes de bocage ? In *Bocages et Sociétés*, Antoine, A. and Marguerie, D. (eds). PUR, Rennes.

Montgomery, D.R. and MacDonald, L.H. (2002). Diagnostic approach to stream channel assessment and monitoring. *Journal of the American Water Resources Association*, 38, 1–16.

Montgomery, D.R., Buffington, J.M., Smith, R.D., Schmidt, K.M., Pess, G. (1995). Pool spacing in forest channels. *Water Resources Research*, 31, 1097–1105.

Moreau, C. (2019). Mettre en débat l'état de référence. Analyse des représentations des dynamiques paysagères au prisme des services écosystémiques. PhD Thesis, University of Montpellier, Montpellier.

Moreau, C., Barnaud, C., Mathevet, R. (2019a). L'évolution des paysages de référence, un angle mort dans la gouvernance des paysages ? *Développement durable et territoires*, 10(2) [Online]. Available at: http://journals.openedition.org/developpementdurable/14341 [Accessed 26 January 2021].

Moreau, C., Barnaud, C., Mathevet, R. (2019b). Conciliate agriculture with landscape and biodiversity conservation: A role-playing game to explore trade-offs among ecosystem services through social learning. *Sustainability*, 11(2), 310.

Moreno-Mateos, D., Maris, V., Bechet, A., Curran, M. (2015). The true loss caused by biodiversity offsets. *Biological Conservation*, 192, 552–559.

Moriceau, J.M. (2011). *L'homme contre le loup : une guerre de deux milles ans*. Fayard, Paris.

Morin, E. (1990). *Introduction à la pensée complexe*. ESF Editeur, Paris.

Morizot, B. (2019). *Raviver les braises du vivant*. Actes Sud, Arles.

Mortier, F. (1990). Sylvigénèse et structures spatiales en forêts tempérées. Synthèse bibliographique et étude de cas : forêt primaire naturelle de plaine, forêt secondaire non exploitée de montagne. Report, DEA, ENGREF, Nancy.

Mosley, G. (1992). La protection de la *wilderness* dans les Alpes australiennes. *Revue de Géographie Alpine*, 80, 307–333.

Moss, B. (2015). Mammals, freshwater reference states, and the mitigation of climate change. *Freshwater Biology*, 60, 1964–1976.

Mountford, E.P. (2002). Fallen dead wood levels in the near natural beech forest at la tillaie reserve, Fontainebleau, France. *Forestry*, 75(2), 203–208.

Muldrow, M., Parsons, E.C., Jonas, R. (2020). Shifting baseline syndrome among coral reef scientists. *Humanities and Social Sciences Communications*, 7(1), 1–8.

Müller, N. (1995). River dynamics and floodplain vegetation and their alterations due to human impact. *Archiv fur Hydrobiologie. Supplementband*, 101, 477–512.

Müller, A., Bøcher, P.K., Svenning, J.C. (2015). Where are the wilder parts of anthropogenic landscapes? A mapping case study for Denmark. *Landscape and Urban Planning*, 144, 90–102.

Murcia, C., Aronson, J., Kattan, G.H., Moreno-Mateos, D., Dixon, K., Simberloff, D. (2014). A critique of the "novel ecosystem" concept. *Trends in Ecology & Evolution*, 29(10), 548–553.

Nash, R. (1993). *Wilderness and the American Mind*. Yale University Press, New Haven.

National Research Council (1992). *Restoration of Aquatic Ecosystems: Science, Technology, and Public Policy*. National Academies Press, Washington, DC.

National Research Council (2002). *Riparian Areas: Functions and Strategies for Management*. National Academies Press, Washington, DC.

Navarro, L.M. and Pereira, H.M. (2012). Rewilding abandoned landscapes in Europe. *Ecosystems*, 15(6), 900–912.

Nogues-Bravo, D., Simberloff, D., Rahbek, C., Sanders, N.J. (2016). Rewilding is the new Pandora's box in conservation. *Current Biology*, 26, R87–R91.

Novacek, M.J. (2001). *The Biodiversity Crisis: Losing What Counts*. The New Press, New York.

Ode, P.R., Rehn, A.C., Mazor, R.D., Schiff, K.C., Stein, E.D., May, J.T., Brown, L.R., Herbst, D.B., Gillett, D., Lunde, K. *et al.* (2014). Evaluating the adequacy of a reference-site pool for ecological assessments in environmentally complex regions. *Freshwater Science*, 35, 237–248.

Ogg, J.G. (2004). Status of divisions of the international geologic time scale. *Lethaia*, 37, 183–199.

Oldeman, R. (1974). Ecotopes des arbres et gradients écologiques verticaux en forêt guyanaise. *Revue d'écologie – la Terre et la Vie*, 218, 487–520.

Oldeman, R. (1990). *Forest, Elements of Sylvology*. Springer-Verlag, Berlin.

Olden, J.D. and Rooney, T.P. (2006). On defining and quantifying biotic homogenization. *Global Ecology and Biogeography*, 15, 113–120.

Omernik, J.M. (1987). Ecoregions of the conterminous United States. *Annals of the Association of American Geographers*, 77, 118–125.

ONCEA (2014). Panorama de la quantification de l'évolution nationale des surfaces agricoles, 29 [Online]. Available at: https://www.viepublique.fr/sites/default/files/rapport/pdf/144000288.pdf.

ONF (1998). Réserves biologiques intégrales, instruction du 30 décembre 1998.

Orsi, F., Geneletti, D., Borsdorf, A. (2013). Mapping wildness for protected area management: A methodological approach and application to the Dolomites UNESCO World Heritage Site (Italy). *Landscape and Urban Planning*, 120, 1–15.

Osborn, G., McCarthy, D., LaBrie, A., Burke, R. (2015). Lichenometric dating: Science or pseudo-science? *Quaternary Research*, 83, 1–12.

Owen-Smith, N. (1987). Pleistocene extinctions – The pivotal role of megaherbivores. *Paleobiology*, 13, 351–362.

Paegelow, M., Camacho Olmedo, M.T., Mas, J.F., Houet, T. (2014). Benchmarking of LUCC modelling tools by various validation techniques and error analysis. *Cybergeo*, 701 [Online]. Available at: http://cybergeo.revues.org/26610.

Paillet, Y., Pernot, C., Boulanger, V., Debaive, N., Fuhr, M., Gilg, O., Gosselin, F. (2015). Quantifying the recovery of old-growth attributes in forest reserves: A first reference for France. *Forest Ecology and Management*, 346, 51–64.

Paine, R. (1980). Food webs – Linkage, interaction strength and community infrastructure – The 3rd Tansley lecture. *Journal of Animal Ecology*, 49, 667–685.

Pala, M., Molina, J., Andreu-Boussut, V. (eds) (2017). *La réserve naturelle régionale de Sainte-Lucie*. Cahiers du Parc naturel régional de la Narbonnaise, 18, Narbonne.

Palierne, J.-M. (1976). Bocages mimétiques d'intercalation et de substitution : genèse et écologie. In *Les bocages : histoire, écologie, économie. Table ronde CNRS : aspects physiques, biologiques et humains des écosystèmes bocagers des régions tempérées humides*. INRA, ENSA, 69–73.

Papworth, S.K., Rist, J., Coad, L., Milner-Gulland, E.J. (2009). Evidence for shifting baseline syndrome in conservation. *Conservation Letters*, 2, 93–100.

Pardo, I., Gómez-Rodríguez, C., Wasson, J.-G., Owen, R., van de Bund, W., Kelly, M., Bennett, C., Birk, S., Buffagni, A., Erba, S. *et al.* (2012). The European reference condition concept: A scientific and technical approach to identify minimally-impacted river ecosystems. *Science of the Total Environment*, 420, 33–42.

Parsons, D.M., Morrison, M.A., MacDiarmid, A.B., Stirling, B., Cleaver, P., Smith, I.W., Butcher, M. (2009). Risks of shifting baselines highlighted by anecdotal accounts of New Zealand's snapper (*Pagrus auratus*) fishery. *New Zealand Journal of Marine and Freshwater Research*, 43(4), 965–983.

Pärtel M., Szava-Kovats R., Zobel M. (2011). Dark diversity: Shedding light on absent species. *Trends in Ecology and Evolution*, 26, 124–128.

Pascal, M., Lorvelec, O., Vigne, J.D. (2006). *Invasions biologiques et extinctions : 11000 ans d'histoire des vertébrés en France*. Editions Quae, Belin.

Pastoureau, M. (2015). *L'ours, histoire d'un roi déchu*. Points, Paris.

Pauly, D. (1995). Anecdotes and the shifting baseline syndrome in fisheries. *Trends in Ecology and Evolution*, 10, 430.

Pereira, H.M. and Navarro, L.M. (eds) (2015). *Rewilding European Landscapes*. Springer International Publishing, New York.

Perigord, M. (1994). Friches et landes en Limousin. *Norois*, 164(1), 611–626.

Perpiña Castillo, C., Coll Aliaga, E., Lavalle, C., Martínez Llario, J.C. (2020). An assessment and spatial modelling of agricultural land abandonment in Spain (2015–2030). *Sustainability*, 12(2), 560.

Pertierra, L.R., Hughes, K.A., Vega, G.C., Olalla-Tárraga, M.A. (2017). High resolution spatial mapping of human footprint across Antarctica and its implications for the strategic conservation of avifauna. *PloS One*, 12, e0168280.

Petts, G.E., Möller, H., Roux, A.L. (eds) (1989). *Historical Change of Large Alluvial Rivers: Western Europe*. Wiley, Chichester, New York.

Peyrache-Gadeau, V. and Perron, L. (2010). Le paysage comme ressource dans les projets de développement territorial. *Développement durable et territoires*, 1(2)

Pic, J., Guen, Y.L., Cossart, É., Fressard, M. (2020). La dynamique d'enfrichement, ses expressions paysagères et ses significations : le cas du Beaujolais viticole. *VertigO*, 5 October.

Pickett, S.T.A. (2013). The flux of nature: Changing worldviews and inclusive concept. In *Linking Ecology and Ethics for a Changing World*, Rozzi, R., Pickett, S.T.A., Palmer, C., Armesto, J.J., Callicott, J.B. (eds). Springer, Cham.

Pinnegar, J.K. and Engelhard, G.H. (2008). The "shifting baseline" phenomenon: A global perspective. *Reviews in Fish Biology and Fisheries*, 18(1), 1–16.

Pioch, S., Barnaud, G., Coïc, B. (2015). Méthodes de dimensionnement des mesures compensatoires pour les zones humides. In *Restaurer la nature pour atténuer les impacts du développement : analyse des mesures compensatoires pour la biodiversité*, Levrel, H., Frascaria-Lacoste, N., Hay, J., Martin, G., Pioch, S. (eds). Editions Quae, Versailles.

Piveteau, V. (1995). *Prospective et territoires : apports d'une réflexion sur le jeu*. Cemagref Editions, Antony.

PNR Alpilles (2015). *Elaboration d'un plan de gestion des haies dans le cadre du programme "LIFE des Alpilles" – Caractérisation écologique des réseaux de haie.* Publication PNR, Saint-Rémy-de-Provence.

Pointereau, P., Coulon, F., Girard, P., Lambotte, M., Stuczynski, T., Ortega, V.S., Anguiano, E., Bamps, C., Terres, J.-M. (2008). Analysis of farmland abandonment and the extent and location of agricultural areas that are actually abandoned or are in risk to be abandoned. Technical report, JRC Publications Repository.

Poirier, L. (1934). Bocage et plaine dans le sud de l'Anjou. *Annales de géographie*, 241, 22–31.

Pollock, M.M., Beechie, T.J., Imaki, H. (2012). Using reference conditions in ecosystem restoration: An example for riparian conifer forests in the Pacific Northwest. *Ecosphere*, 3, 1–23.

Pontailler, J.Y., Faille, A., Lemée, G. (1997). Storms drive successional dynamics in natural forests: A case study in Fontainebleau forest (France). *Forest Ecology and Management*, 98(1), 1–15.

Potapov, P.V., Yaroshenko, A.Y., Turubanova, S., Dubinin, M., Laestadius, L., Thies, C., Aksenov, D., Egorov, A., Yesipova, Y., Glushkov, I. *et al.* (2008). Mapping the world's intact forest landscapes by remote sensing. *Ecology and Society*, 13, 51.

Poussou, J.P. (1999). *La Terre et les Paysans en France et en Grande-Bretagne aux XVIIe et XVIIIe siècles*. CNED/SEDES, Paris.

Poux, X. (2003). Les méthodes des scénarios. In *Prospectives pour l'environnement : quelles recherches ? Quelles ressources ? Quelles méthodes ?*, Mermet L. (ed.). La Documentation Française, Paris.

Poux, X., Narcy, J.B., Ramain, B. (2009). Réinvestir le saltus dans la pensée agronomique moderne : vers un nouveau front éco-politique ? *L'Espace Politique*, 9 [Online]. Available at: http://journals.openedition.org/espacepolitique/1495 [Accessed 15 January 2021].

Preux, T., Augier, A., Laslier, M., Dufour, S., Magueur, A. (2019). L'enfrichement des milieux humides en Bretagne : dynamique paysagère, évolution des usages, effets sur l'écologie des milieux aquatiques. Scientific report, ECOFRICHE phase 1 program, FMA, Université Rennes 2, LETG-Rennes.

Price, A.R.G., Ghazi, S.J., Tkaczynski, P.J., Venkatachalam, A.J., Santillan, A., Pancho, T., Metcalfe, R., Saunders, J. (2014). Shifting environmental baselines in the Red Sea. *Marine Pollution Bulletin*, 78(1–2), 96–101.

Price, B., Kienast, F., Seidl, I., Ginzler, C., Verburg, P.H., Bolliger, J. (2015). Future landscapes of Switzerland: Risk areas for urbanisation and land abandonment. *Applied Geography*, 57, 32–41.

Quétier, F., Regnery, B., Levrel, H. (2014). No net loss of biodiversity or paper offsets? A critical review of the French no net loss policy. *Environmental Science & Policy*, 38, 120–131.

Rameau, J. (1987). Contribution phytoécologique et dynamique à l'étude des écosystèmes. Applications aux forêts du Nord-Est de la France. PhD Thesis, Université de Franche-Comté, Besançon.

Rameau, J. (1995). Définition et évaluation de la qualité d'un écosystème forestier. *Forêt Wallonne*, 39/40, 38–51.

Raven, P.J., Holmes, N.T.H., Charrier, P., Dawson, F.H., Naura, M., Boon, P.J. (2002). Towards a harmonized approach for hydromorphological assessment of rivers in Europe: A qualitative comparison of three survey methods. *Aquatic Conservation: Marine and Freshwater Ecosystems*, 12, 405–424.

Regnery, B., Quétier, F., Cozannet, N., Gaucherand, S., Laroche, A., Burylo, M., Couvet, D., Kerbiriou, C. (2013). Mesures compensatoires pour la biodiversité : comment améliorer les dossiers environnementaux et la gouvernance ? *Sciences eaux & territoires*, 12, 1–8.

Reise, K. (2013). *A Natural History of the Wadden Sea. Riddled by Contingencies*. Waddenacademie, Common Wadden Sea Secretariat, Leeuwarden, Wilhelmshaven.

Rey Benayas, J. (2019). Rewilding: As farmland and villages are abandoned, forests, wolves and bears are returning to Europe. *The Conversation* [Online]. Available at: https://theconversation.com/rewilding-as-farmland-and-villages-are-abandoned-forests-wolves-and-bears-are-returning-to-europe-119316.

Reynaud, D. (2018). L'agriculture française en Europe de 2000 à 2015. Au premier rang avec cependant une érosion de ses positions. *INSEE première*, 1704.

Rheinhardt, R., Brinson, M., Brooks, R., McKenney-Easterling, M., Rubbo, J.M., Hite, J., Armstrong, B. (2007). Development of a reference-based method for identifying and scoring indicators of condition for coastal plain riparian reaches. *Ecological Indicators*, 7, 339–361.

Ridder, B. (2007). The naturalness versus wildness debate: Ambiguity, inconsistency, and unattainable objectivity. *Restoration Ecology*, 15(1), 8–12.

Riggio, J., Baillie, J.E.M., Brumby, S., Ellis, E., Kennedy, C.M., Oakleaf, J.R., Tait, A., Tepe, T., Theobald, D.M., Venter, O. *et al.* (2020). Global human influence maps reveal clear opportunities in conserving Earth's remaining intact terrestrial ecosystems. *Global Change Biology*, 26, 4344–4356.

Ripple, W.J. and Beschta, R.L. (2012). Trophic cascade in Yellowstone: The first 15 years after wolf reintroduction. *Biological Conservation*, 145, 205–213.

Ritchie, E.G., Elmhagen, B., Glen, A.S., Letnic, M., Ludwig, G., McDonald, R.A. (2012). Ecosystem restoration with teeth: What role for predators? *Trends in Ecology and Evolution*, 27, 265–271.

Roberts, C.N., Fyfe, R.M., Woodbridge, J., Gaillard, M.-J., Davis, B.A.S., Kapla, J.O., Marquer, L., Mazier, F., Nielsen, A.B., Sugita, S. *et al.* (2018). Europe's lost forests: A pollen-based synthesis for the last 11,000 years. *Scientific Reports*, 8, 1–8.

Roche, P.A., Billen, G., Bravard, J.-P., Décamps, H., Pennequin, D., Vindimian, E., Wasson, J.-G. (2005). Les enjeux de recherche liés à la directive-cadre européenne sur l'eau. *Comptes rendus géoscience*, 337, 243–267.

Rodriguez, J., Rodriguez-Clark, K., Baillie, J., Neville, A., Benson, J., Boucher, T., Brown, C., Burgess, N., Collen, B., Jennings, M. *et al.* (2011). Establishing IUCN red list criteria for threatened ecosystems. *Conservation Biology*, 25(1), 21–29.

Rosgen, D.L. (1994). A classification of natural rivers. *CATENA*, 22, 169–199.

Rotherham, I.D. (2014). The call of the wild. Perceptions, history people and ecology in the emerging paradigms of wilding. *ECOS*, 35, 35–43.

Rotherham, I.D. (2015). Wild in the eye of the beholder. *Conservation Land Management*, 13, 12–13.

Rouveyrol, P., Léonard, L., Herard, K. (2019). Note sur les concepts de "pleine naturalité" et de "protection forte" pour soutenir la définition de nouveaux objectifs dans le cadre de la Stratégie nationale aires protégées 2020–2030. Technical report, UMS Patrimoine Naturel.

Rubenstein, D.R., Rubenstein, D.I., Sherman, P.W., Gavin, T.A. (2006). Pleistocene park: Does re-wilding North America represent sound conservation for the 21st century? *Biological Conservation*, 132, 232–238.

Ruddiman, W.F. (2003). The anthropogenic greenhouse era began thousands of years ago. *Climatic Change*, 61, 261–293.

Ruddiman, W.F., He, F., Vavrus, F.J., Kutzbach, J.E. (2020). The early anthropogenic hypothesis: A review. *Quaternary Science Reviews*, 240, 106386.

Russel, J.A., Matthews, J.H., Jones, R. (1979). Wilderness in Tasmania: A report to the Australian heritage commission. Occasional Paper 10, Centre for Environmental Studies, University of Tasmania, Hobart.

Rutz, C., Loretto, M.-C., Bates, A.E., Davidson, S.C., Duarte, C.M., Jetz, W., Johnson, M., Kato, A., Kays, R., Mueller, T. *et al.* (2020). COVID-19 lockdown allows researchers to quantify the effects of human activity on wildlife. *Nature Ecology & Evolution*, 4(9), 1156–1159.

Sabatini, F.M., Burrascano, S., Keeton, W.S., Levers, C., Lindner, M., Pötzschner, F., Verkerk, P.J., Bauhus, J., Buchwald, E., Chaskovsky, O. *et al.* (2018). Where are Europe's last primary forests? *Diversity and Distributions*, 24, 1426–1439.

Salingue, Y. (2015). *La quête de l'ours*. Editions du Rouergue, Rodez.

Sanderson, E.W., Jaiteh, M., Levy, M.A., Redford, K.H., Wannebo, A.V., Woolmer, G. (2002). The human footprint and the last of the wild. *BioScience*, 52, 891–904.

Sauer, C.O. (1925). The morphology of landscape. In *Changing Approaches to Landscapes*, Ingerson, A.E. (ed.). Institute for Cultural Landscape Studies, Harvard.

Schletterer, M., Füreder, L., Kuzovlev, V.V., Zhenikov, Y.N., Grigorieva, I.L. (2014). Lowland river reference condition: Spatial and temporal patterns of the zoobenthos community in the Volga headwaters (2006–2010). *Hydrobiologia*, 729, 175–189.

Schlosberg, D. (2017). La gestion environnementale à l'heure de l'Anthropocène. *La Pensée écologique*, 1 [Online]. Available at: https://www.cairn.info/revue-la-pensee-ecologique-2017-1-page-f.htm [Accessed 15 January 2021].

Schmerber, J.M. (1950). *La réorganisation foncière en France : le remembrement rural*. La maison rustique, Strasbourg.

Schnitzler, A. (2014). Towards a new European wilderness: Embracing unmanaged forest growth and the decolonisation of nature. *Landscape and Urban Planning*, 126, 74–80.

Schnitzler, A. and Génot, J.-C. (2012). *La France des friches : de la ruralité à la féralité* Éditions Quae, Versailles.

Schnitzler, A., Génot, J.-C., Wintz, M. (2008a). Espaces protégés : de la gestion conservatoire vers la non-intervention. *Courrier de l'Environnement de l'INRA*, 56, 29–43.

Schnitzler, A., Génot, J.-C., Wintz, M., Hale, B.W. (2008b). Naturalness and conservation in France. *Journal of Agricultural and Environmental Ethics*, 21(5), 423–436.

Schnitzler-Lenoble, A. (2002). *Écologie des forêts naturelles d'Europe. Biodiversité, sylvigénèse, valeur patrimoniale des forêts primaires*. Tec & Doc, Lavoisier, Paris.

Scholes, R.J. and Biggs, R. (2005). A biodiversity intactness index. *Nature*, 434, 45–49.

Seddon, P.J., Griffiths, C.J., Soorae, P.S., Armstrong, D.P. (2014). Reversing defaunation: Restoring species in a changing world. *Science*, 345, 406–412.

Sentilles, J., Lemaitre, P.L., Vanpe, C., Quenette, P.Y. (2020). Rapport annuel du Réseau Ours Brun 2019. Report, Office Français de la Biodiversité.

Shakun, J., Clark, P., He, F., Marcott, S., Mix, A., Liu, Z., Otto-Bliesner, B., Schmittner, A., Bard, E. (2012). Global warming preceded by increasing carbon dioxide concentrations during the last deglaciation. *Nature*, 484, 49–54.

Shinneman, D.J., Baker, W.L., Lyon, P. (2008). Ecological restoration needs derived from reference conditions for a semi-arid landscape in Western Colorado, USA. *Journal of Arid Environments*, 72, 207–227.

Simmonds, J.S., Sonter, L.J., Watson, J.E., Bennun, L., Costa, H.M., Dutson, G., Edwards, S., Grantham, H., Griffiths, V.F., Jones, J.P.G. *et al.* (2020). Moving from biodiversity offsets to a target-based approach for ecological compensation. *Conservation Letters*, 13:e12695.

Simon, L. (1999). Potentialités forestières et production de bois : de l'approche naturaliste à l'approche géographique. *Annales de géographie*, 635–647.

Singarayer, J.S., Valdes, P.J., Friedlingstein, P., Nelson, S., Beerling, D.J. (2011). Late Holocene methane rise caused by orbitally controlled increase in tropical sources. *Nature*, 470, 82–85.

Slaymaker, O. (1991). Mountain geomorphology: A theoretical framework for measurement programmes. *Catena*, 18(5), 427–437.

Smith, C.I. (2005). Re-wilding: Introductions could reduce biodiversity. *Nature*, 437, 318.

Smith, B. and Zeder, M. (2013). The onset of the Anthropocene. *Anthropocene,* 4, 8–13.

Soar, P.J. and Thorne, C.R. (2001). Channel restoration design for meandering rivers. Coastal and Hydraulics Laboratory, US Army Engineer Research and Development Center, Vicksburg.

Society for Ecological Restoration International Science & Policy Working Group (2004). The SER international primer on ecological restoration. Society for Ecological Restoration International, Tucson.

Society for Ecological Restoration Science and Policy Working Group (2002). The SER primer on ecological restoration. Society for Ecological Restoration.

Soga, M. and Gaston, K.J. (2018). Shifting baseline syndrome: Causes, consequences, and implications. *Frontiers in Ecology and the Environment*, 16(4), 222–230.

Sonter, L.J., Simmonds, J.S., Watson, J.E.M., Jones, J.P.G., Kiesecker, J.M., Costa, H.M., Bennun, L., Edwards, S., Grantham, H.S., Griffiths, V.F. *et al.* (2020). Local conditions and policy design determine whether ecological compensation can achieve No Net Loss goals. *Nature Communications*, 11, 2072.

Soulé, M.E. (1986). *Conservation Biology. The Science of Scarcity and Diversity*. Sinauer Associates, Sunderland.

Soulé, M.E. (1990). The onslaught of alien species, and other challenges in the coming decades. *Conservation Biology*, 4, 233–240.

Soulé, M.E. and Noss, R. (1998). Rewilding and biodiversity: Complementary goals for continental conservation. *Wild Earth*, 8(3), 18–28.

Soulé, M.E. and Wilcox, B.A. (eds) (1980). *Conservation Biology: An Evolutionary-Ecological Perspective*. Sinauer Associates, Sunderland.

Spash, C.L. (2015). Bulldozing biodiversity: The economics of offsets and trading-in nature. *Biological Conservation*, 192, 541–551.

Speer, J.H. (2011). *Fundamentals of Tree-Ring Research.* University of Arizona Press, Tucson.

Stanton, J.P. and Morgan, M.G. (1977). Project "RAKES": A rapid appraisal of key and endangered sites, the Queensland case study. University of New England, Armidae.

Steffen, W., Sanderson, A., Tyson, P.D., Jäger, J., Matson, P.A., Moore III, B., Oldfield, F., Richardson, K., Schellnhuber, H.J., Turner II, B.L. *et al.* (2004). *Global Change and the Earth System: A Planet Under Pressure. Executive Summary.* Springer-Verlag, New York.

Steffen, W., Crutzen, P.J., McNeill, J.R. (2007). The Anthropocene: Are humans now overwhelming the great forces of nature? *Ambio*, 36, 614–621.

Stuart, A.J. (2015). Late quaternary megafaunal extinctions on the continents: A short review. *Geological Journal*, 50, 338–363.

Sun, Z. and Ren, H. (2011). Ecological memory and its potential applications in ecology: A review. *The Journal of Applied Ecology*, 22, 549–55.

Surian, N. and Rinaldi, M. (2004). Channel adjustments in response to human alteration of sediment fluxes: Examples from Italian rivers. In *Sediment Transfer through the Fluvial System*, Golosov, V., Belyaev, V., Walling, D.E. (eds). IAHS Publication, Wallingford.

Sutherland, D.R., Glen, A.S., de Tores, P.J. (2011). Could controlling mammalian carnivores lead to mesopredator release of carnivorous reptiles? *Proceedings of the Royal Society B*, 278, 641–648.

Svenning, J.S., Perdersen, P.B.M., Donlan, J., Erjnaes, R., Faurby, S., Galetti, M., Hansen, D.M., Sandel, B., Sandom, C.J., Terborgh, J.W. *et al.* (2016). Science for a wilder Anthropocene: Synthesis and future directions for trophic rewilding research. *Proceedings of the National Academy of Sciences*, 113, 898–906.

Szabo, J.K., Khwaja, N., Garnett, S.T., Butchart, S.H.M. (2012). Global patterns and drivers of avian extinctions at the species and subspecies level. *PLoS One*, October 8. DOI: 10.1371/journal.pone.0047080.

Szczypta, C., Gascoin, S., Houet, T., Vigneau, C., Fanise, P. (2015). Impact of climate and land-use changes on snow cover in a small headwater catchment in the Pyrenees (France). *Journal of Hydrology*, 521, 84–99.

Taccoen, A., Piedallu, C., Seynave, I., Perez, V., Gegout-Petit, A., Nageleisen, L.-M., Bontemps, J.-D., Gegout, J.-C. (2019). Background mortality drivers of European tree species: Climate change matters. *Proceedings of the Royal Society B*, 286, 1900.

Tansley, A.G. (1935). The use and abuse of vegetational concepts and terms. *Ecology*, 16, 284–307.

Tarabon, S., Calvet, C., Delbar, V., Isselin, F., Dutoit, T. (2020). Integrating a landscape connectivity approach to the mitigation hierarchy planning following urbanization dynamics. *Landscape and Urban Planning*, 202, 103871.

Thoreau, H.D. (1854). *Walden; or Life in the Woods*. Ticknor and Fields, Boston.

Tree, I. (2019). *Wilding: The Return of Nature to a British Farm*. Picador, London.

Tricker, J. and Landres, P. (2018). Mapping threats to wilderness character in the National Wilderness Preservation System. *Biological Conservation*, 227, 243–251.

Tsing, A. (2017). The buck, the bull, and the dream of the stag: Some unexpected weeds of the Anthropocene. *Suomen Anthropologi*, 42(1), 19.

Turner, W.R., Nakamura, T., Dinetti, M. (2004). Global urbanization and the separation of humans from nature. *Bioscience*, 54(6), 585–590.

Turvey, S.T. (2009). Holocene mammal extinctions. In *Holocene Extinctions*, Turvey, S.T. (ed.). Oxford University Press, Oxford.

Turvey, S.T., Barrett, L.A., Yujiang, H.A.O., Lei, Z., Xinqiao, Z., Xianyan, W., Ding, W. (2010). Rapidly shifting baselines in Yangtze fishing communities and local memory of extinct species. *Conservation Biology*, 24(3), 778–787.

Vacquié, L., Houet, T., Sohl, T., Reeker, R., Sayler, K. (2015). Developing scenarios to project LULC changes in the Pyrenees (France): A model-based approach to assess land abandonment and reforestation dynamics. *Journal of Mountain Science*, 12(4), 905–920.

Vacquié, L., Houet, T., Sheeren, D., de Munnik, N., Roussel, V., Waddle, J. (2016). Adapting grazing practices to limit the reforestation of mountainous summer pastures: An agent-based model (SHIELD). *Environmental Modelling and Software*, 84, 395–411.

Vaissière, A.C., Quétier, F., Calvet, C., Levrel, H., Wunder, S. (2020). Biodiversity offsets and payments for environmental services: Clarifying the family ties. *Ecological Economics*, 169, 106–428.

Vallauri, D., André, J., Génot, J.-C., Palma, P., Eynard-Machet, R. (2010). *Biodiversité, naturalité, humanité. Pour inspirer la gestion des forêts*. Lavoisier, Paris.

Van Der Zanden, E.H., Verburg, P.H., Schulp, C.J.E., Verkerk, P.J. (2017). Trade-offs of European agricultural abandonment. *Land Use Policy*, 62, 290–301.

Van Strien, A.J., van Swaay C.A.M., van Strien-van Liempt W.T.F.H., Poot, M.J.M., WallisDeVriesb, M.F. (2019). Over a century of data reveal more than 80% decline in butterflies in the Netherlands. *Biological Conservation*, 234, 116–122.

Vartanyan, S.L.V., Garutt, V.E., Sher, A.V. (1993). Holocene dwarf mammoths from Wrangel Island in the Siberian Arctic. *Nature*, 362, 337–340.

Venter, O., Sanderson, E.W., Magrach, A., Allan, J.R., Beher, J., Jones, K.R., Possingham, H.P., Laurance, W.F., Wood, P., Fekete, B.M. *et al.* (2016). Sixteen years of change in the global terrestrial human footprint and implications for biodiversity conservation. *Nature Communications*, 7, 12558.

Vera, F. (2000). *Grazing Ecology and Forest History*. Caby Publishing, Wallingford.

Vera, F. (2009). Large-scale nature development. The Oostvaardersplassen. *British Wildlife*, June, 28–36.

Vera, F. (2010). The shifting baseline syndrome in restoration ecology. *Restoration and History: The Search for a Usable Environmental Past*, 98–110.

Vera, F. (2018). Rewilding. Is it nature conservation? [Online]. Available at: http://naturmoedet.dk/wp-content/uploads/2018/06/Frans-Vera-Rewilding.pdf.

Vigne, J.D. (1988). *Les mammifères post-glaciaires de Corse. Etude archéozoologique.* Editions du CNRS, Paris.

Vigne, J.D. (1992). Zooarchaeology and biogeographical history of the mammals of Corsica and Sardinia since the last ice-age. *Mammal Review*, 22, 87–96.

Vigne, J.D. (2004). *Les débuts de l'élevage*. Le Pommier, Paris.

Vigne, J.D. and Valladas, H. (1996). Small mammal fossil assemblages as indicators of environmental change in northern Corsica during the last 2500 years. *Journal of Archaeological Science*, 23, 199–215.

Voiron, C. (2006) L'espace dans la modélisation des interactions nature-société. Paper, Actes du colloque international interactions nature-société, analyse et modèles, La Baule, UMR 6554 LETG, 3–6 May.

Von Witzke, H. and Noleppa, S. (2010). EU agricultural production and trade: Can more efficiency prevent increasing "land-grabbing" outside of Europe? Study, OPERA.

Wackernagel, M. and Rees, W. (1996). What is an ecological footprint? In *The Sustainable Urban Development Reader*, Wheeler, S.M. and Beatley, T. (eds). Routledge, New York.

Wallin, M., Wiederholm, T., Johnson, R.K. (2003). Guidance on establishing reference conditions and ecological status class boundaries for inland surface waters. Report, REFCOND.

Wapner, P. (2010). *Living through the End of Nature: The Future of American Environmentalism*. MIT Press, Cambridge.

Ward, J.V., Tockner, K., Edwards, P.J., Kollmann, J., Bretschko, G., Gurnell, A.M., Petts, G.E., Rossaro, B. (1999). A reference river system for the Alps: The Fiume Tagliamento. *Regulated Rivers: Research & Management*, 15, 63–75.

Wasson, J.-G., Chandesris, A., Pella, H., Blanc, L. (2002). *Les hydro-écorégions de France métropolitaine. Approche régionale de la typologie des eaux courantes et éléments pour la définition des peuplements de référence d'invertébrés.* Ministère de l'Ecologie et du Développement Durable, Cemagref BEA/LHQ.

Waters, C.N., Zalasiewicz, J., Summerhayes, C., Barnosky, A.D., Poirier, C., Gakuszka, A. (2016). The Anthropocene is functionally and stratigraphically distinct from the Holocene. *Science*, 351, 6269.

Watson, J.E.M., Shanahan, D.F., Di Marco, M., Allan, J., Laurance, W.F., Sanderson, E.W., Mackey, B., Venter, O. (2016). Catastrophic declines in wilderness areas undermine global environment targets. *Current Biology*, 26, 2929–2934.

Watt, A.S. (1947). Pattern and process in the plant community. *Journal of Ecology*, 35, 1–22.

Watteaux, M. (2012). Le bocage. Un paysage rural à la lumière des études archéologiques et archéogéographiques. *Archéopages*, 34, 64–72.

Weber, E. (1983). *La Fin des terroirs. La modernisation de la France rurale, 1870–1914*. Fayard Ed. Recherches, Paris.

Weinstock, J., Willerslev, E., Sher, A.V., Tong, W., Ho, S.Y.W., Rubenstein, D.I., Storer, J., Burns, A., Martin, L.D., Bravi, C. *et al.* (2005). Evolution, systematics, and phylogeography of Pleistocene horses in the new world: A molecular perspective. *PLoS Biology*, 3, 1373–1379.

Weissgerber, M., Roturier, S., Julliard, R., Guillet, F. (2019). Biodiversity offsetting: Certainty of the net loss but uncertainty of the net gain. *Biological Conservation*, 237, 200–208.

Whatmore, S. (2002). *Hybrid Geographies. Natures, Cultures, Spaces*. SAGE Publications, London.

Whipple, A.A., Grossinger, R.M., Davis, F.W. (2011). Shifting baselines in a California oak savanna: Nineteenth century data to inform restoration scenarios. *Restoration Ecology*, 19(101), 88–101.

White, P.S. and Walker, J.L. (1997). Approximating nature's variation: Selecting and using reference information in restoration ecology. *Restoration Ecology*, 5, 338–349.

Whittier, T.R., Stoddard, J.L., Larsen, D.P., Herlihy, A.T. (2007). Selecting reference sites for stream biological assessments: Best professional judgment or objective criteria. *Journal of the North American Benthological Society*, 26, 349–360.

Wilson, E.O. (2012). *Biophilie*. Corti, Paris.

Wilson, D.E. and Mittermeier, R.A. (2009). *Handbook of the Mammals of the World. Vol. 1. Carnivores*. Lynx Edicions, Barcelona.

Winter, S. (2012). Forest naturalness assessment as a component of biodiversity monitoring and conservation management. *Forestry*, 4, 1–12.

Winter, S., Fischer, H.S., Fischer, A. (2010). Relative quantitative reference approach for naturalness assessments of forests. *Forest Ecology and Management*, 259(8), 1624–1632.

Wolfe, A.P., Hobbs, W.O., Birks, H.H., Briner, J.P., Holmgren, S.U., Ingólfsson, O., Kaushal, S.S., Miller, G.H., Pagani, M., Saros, J.E. *et al.* (2013). Stratigraphic expressions of the Holocene-Anthropocene transition revealed in sediments from remote lakes. *Earth-Science Reviews*, 116, 17–34.

Woolmer, G., Trombulak, S.T., Ray, J.C., Doran, P.J., Anderson, M.G., Baldwin, R.F., Morgan, A., Sanderson, E.W. (2008). Rescaling the human footprint: A tool for conservation planning at an ecoregional scale. *Landscape and Urban Planning*, 87, 42–53.

WWF (2020). *Living Planet Report – 2020: Bending the Curve of Biodiversity Loss*. Almond, R.E.A., Grooten, M., Petersen, T. (eds). WWF, Gland.

Wyżga, B., Zawiejska, J., Radecki-Pawlik, A., Hajdukiewicz, H. (2012). Environmental change, hydromorphological reference conditions and the restoration of Polish Carpathian rivers: Environmental change and reference conditions. *Earth Surface Processes and Landforms*, 37, 1213–1226 [Online]. Available at: https://doi.org/10.1002/esp.3273.

Yang, F., Matsushita, B., Yang, W., Fukushima, T. (2014). Mapping the human footprint from satellite measurements in Japan. *ISPRS Journal of Photogrammetry and Remote Sensing*, 88, 80–90.

Yaroshenko, A.Y., Potapov, P.V., Turubanova, S.A. (2001). *The Last Intact Forest Landscapes of Northern European Russia*. Greenpeace Russia and Global Forest Watch, Moscow.

Zalasiewicz, J., Waters, C.N., Williams, M., Barnosky, A.D., Cearreta, A., Crutzen, P., Ellis, E., Ellis, M.A., Fairchild, I.J., Grinevald J. *et al.* (2015). When did the Anthropocene begin? A mid-twentieth century boundary level is stratigraphically optimal. *Quaternary International*, 383, 196–203.

Zimov, S.A., Chuprynin, V.I., Oreshko, A.P., Chapin III, F.S., Reynolds, J.F., Chapin, M.C. (1995) Steppe-tundra transition: A herbivore-driven biome shift at the end of the Pleistocene. *American Naturalist*, 146, 765–794.

List of Authors

Vincent ANDREU-BOUSSUT
UMR ESO
University of Le Mans
France

Cécile BARNAUD
UMR DYNAFOR
INRAE
France

Coralie CALVET
UMR CEFE
University of Montpellier
France

Jonathan CARRUTHERS-JONES
University of Leeds
UK

Céline CHADENAS
UMR LETG-Nantes
Nantes University
France

Armelle DECAULNE
CNRS
UMR LETG-Nantes
Nantes University
France

Simon DUFOUR
UMR LETG-Rennes
University of Rennes 2
France

Laurent GODET
CNRS
UMR LETG-Nantes
Nantes University
France

Adrien GUETTÉ
UR ADI-Suds
ISTOM
Angers
France

Thomas HOUET
CNRS
UMR LETG-Rennes
University of Rennes 2
France

Damien MARAGE
UMR ThéMA
University of Franche-Comté
Besançon
France

Raphaël MATHEVET
CNRS
UMR CEFE
Montpellier
France

David MONTEMBAULT
UMR ESO
Agrocampus Ouest
Angers
France

Clémence MOREAU
UMR SENS
IRD
Marseille
France

Anne-Julia ROLLET
UMR LETG-Rennes
University of Rennes 2
France

Index

E, F

G, H

I, L

M, N

O, P

R, S

T, U

V, W

Y, Z

Other titles from

iSTE

in

Ecological Science

2022

GRISON Claude, CASES Lucie, LE MOIGNE Mailys, HOSSAERT-MCKEY Martine
Photovoltaism, Agriculture and Ecology From Agrivoltaism to Ecovoltaism

LÉVÊQUE Christian
Biodiversity Erosion: Issues and Questions

2021

FLEURENCE Joël
Microalgae: From Future Food to Cellular Factory

AMIARD Jean-Claude
Disarmament and Decommissioning in the Nuclear Domain (Radioactive Risk Set – Volume 4)

JOLY Fernand, BOURRIÉ Guilhem
Mankind and Deserts 1: Deserts, Aridity, Exploration and Conquests
Mankind and Deserts 2: Water and Salts
Mankind and Deserts 3: Wind in Deserts and Civilizations

MUTTIN Frédéric, THOMAS Hélène
Marine Environmental Quality

PIOCH Sylvain, SOUCHE Jean-Claude
Eco-design of Marine Infrastructures: Towards Ecologically-informed Coastal and Ocean Development

2020

BRUSLÉ Jacques, QUIGNARD Jean-Pierre
Fish Behavior 1: Eco-ethology
Fish Behavior 2: Ethophysiology

LE FLOCH Stéphane
Remote Detection and Maritime Pollution: Chemical Spill Studies

MIGON Christophe, NIVAL Paul, SCIANDRA Antoine
The Mediterranean Sea in the Era of Global Change 1: 30 Years of Multidisciplinary Study of the Ligurian Sea
The Mediterranean Sea in the Era of Global Change 2: 30 Years of Multidisciplinary Study of the Ligurian Sea

ROSSIGNOL Jean-Yves
Climatic Impact of Activities: Methodological Guide for Analysis and Action

2019

AMIARD Jean-Claude
Industrial and Medical Nuclear Accidents: Environmental, Ecological, Health and Socio-economic Consequences
(Radioactive Risk SET – Volume 2)
Nuclear Accidents: Prevention and Management of an Accidental Crisis
(Radioactive Risk SET – Volume 3)

BOULEAU Gabrielle
Politicization of Ecological Issues: From Environmental Forms to Environmental Motives

DAVID Valérie
Statistics in Environmental Sciences

GIRAULT Yves
UNESCO Global Geoparks: Tension Between Territorial Development and Heritage Enhancement

KARA Mohamed Hichem, QUIGNARD Jean-Pierre
Fishes in Lagoons and Estuaries in the Mediterranean 2: Sedentary Fish
Fishes in Lagoons and Estuaries in the Mediterranean 3A: Migratory Fish
Fishes in Lagoons and Estuaries in the Mediterranean 3B: Migratory Fish

OUVRARD Benjamin, STENGER Anne
Incentives and Environmental Policies: From Theory to Empirical Novelties

2018

AMIARD Jean-Claude
Military Nuclear Accidents: Environmental, Ecological, Health and Socio-economic Consequences
(Radioactive Risk SET – Volume 1)

FLIPO Fabrice
The Coming Authoritarian Ecology

GUILLOUX Bleuenn
Marine Genetic Resources, R&D and the Law 1: Complex Objects of Use

KARA Mohamed Hichem, QUIGNARD Jean-Pierre
Fishes in Lagoons and Estuaries in the Mediterranean 1: Diversity, Bioecology and Exploitation

2016

BAGNÈRES Anne-Geneviève, HOSSAERT-MCKEY Martine
Chemical Ecology

2014

DE LARMINAT Philippe
Climate Change: Identification and Projections

Printed and bound by CPI Group (UK) Ltd, Croydon, CR0 4YY
13/09/2026
14961042-0006